高等职业教育系列教材

电子与通信工程专业英语
第 2 版

主　编　徐存善　王存才　席东河
副主编　周志宇　唐红莲　段峰松
参　编　蒋志豪　谭胡心　徐　起

机械工业出版社

本书由电子技术基础、通信技术、计算机技术和高级电子与通信工程4部分组成，共有22个单元。每个单元包括课文、生词与短语、解析、练习、翻译技巧和阅读材料，语言通俗易懂，内容生动新颖，应用性强，有助于从不同侧面有针对性地培养学生的专业英语实用技能，提高学生阅读和翻译英文技术文献与专业资料的能力，以及在一定场景下用英语进行交流的能力。书后附有参考译文和大部分习题答案，便于帮助学生理解和检查自己所掌握的内容，在一定程度上减轻了学习的难度。

本书适合高职高专院校电子类、通信类专业和计算机控制等专业的学生使用，也可供相关专业的工程技术人员参考学习。

本书配有授课电子课件，需要的教师可登录www.cmpedu.com免费注册，审核通过后下载，或联系编辑索取（QQ：1239258369，电话：010-88379739）。

图书在版编目（CIP）数据

电子与通信工程专业英语/徐存善，王存才，席东河主编. —2版. —北京：机械工业出版社，2017.7（2021.8重印）
高等职业教育系列教材
ISBN 978-7-111-57848-2

Ⅰ.①电… Ⅱ.①徐… ②王… ③席… Ⅲ.①电子技术–英语–高等职业教育–教材②通信工程–英语–高等职业教育–教材 Ⅳ.①TN②TN91

中国版本图书馆CIP数据核字（2017）第210741号

机械工业出版社（北京市百万庄大街22号 邮政编码100037）
策划编辑：王 颖 责任编辑：王 颖
责任校对：梁 倩 责任印制：郜 敏
北京富资园科技发展有限公司印刷
2021年8月第2版第3次印刷
184mm×260mm·14.75印张·377千字
标准书号：ISBN 978-7-111-57848-2
定价：39.90元

凡购本书，如有缺页、倒页、脱页，由本社发行部调换

电话服务	网络服务
服务咨询热线：010 – 88379833	机 工 官 网：www.cmpbook.com
读者购书热线：010 – 88379649	机 工 官 博：weibo.com/cmp1952
	教育服务网：www.cmpedu.com
封面无防伪标均为盗版	金 书 网：www.golden-book.com

高等职业教育系列教材
电子类专业编委会成员名单

主　　任　曹建林

副 主 任　（按姓氏笔画排序）

于宝明　王钧铭　任德齐　华永平　刘　松　孙　萍
孙学耕　杨元挺　杨欣斌　吴元凯　吴雪纯　张中洲
张福强　俞　宁　郭　勇　曹　毅　梁永生　董维佳
蒋蒙安　程远东

委　　员　（按姓氏笔画排序）

丁慧洁　王卫兵　王树忠　王新新　牛百齐　吉雪峰
朱小祥　庄海军　关景新　孙　刚　李菊芳　李朝林
李福军　杨打生　杨国华　肖晓琳　何丽梅　余　华
汪赵强　张静之　陈　良　陈子聪　陈东群　陈必群
陈晓文　邵　瑛　季顺宁　郑志勇　赵航涛　赵新宽
胡　钢　胡克满　闾立新　姚建永　聂开俊　贾正松
夏玉果　夏西泉　高　波　高　健　郭　兵　郭雄艺
陶亚雄　黄永定　黄瑞梅　章大钧　商红桃　彭　勇
董春利　程智宾　曾晓宏　詹新生　廉亚因　蔡建军
谭克清　戴红霞　魏　巍　瞿文影

秘 书 长　胡毓坚

出版说明

《国家职业教育改革实施方案》（又称"职教20条"）指出：到2022年，职业院校教学条件基本达标，一大批普通本科高等学校向应用型转变，建设50所高水平高等职业学校和150个骨干专业（群）；建成覆盖大部分行业领域、具有国际先进水平的中国职业教育标准体系；从2019年开始，在职业院校、应用型本科高校启动"学历证书＋若干职业技能等级证书"制度试点（即1＋X证书制度试点）工作。在此背景下，机械工业出版社组织国内80余所职业院校（其中大部分院校入选"双高"计划）的院校领导和骨干教师展开专业和课程建设研讨，以适应新时代职业教育发展要求和教学需求为目标，规划并出版了"高等职业教育系列教材"丛书。

该系列教材以岗位需求为导向，涵盖计算机、电子、自动化和机电等专业，由院校和企业合作开发，多由具有丰富教学经验和实践经验的"双师型"教师编写，并邀请专家审定大纲和审读书稿，致力于打造充分适应新时代职业教育教学模式、满足职业院校教学改革和专业建设需求、体现工学结合特点的精品化教材。

归纳起来，本系列教材具有以下特点：

1）充分体现规划性和系统性。系列教材由机械工业出版社发起，定期组织相关领域专家、院校领导、骨干教师和企业代表召开编委会年会和专业研讨会，在研究专业和课程建设的基础上，规划教材选题，审定教材大纲，组织人员编写，并经专家审核后出版。整个教材开发过程以质量为先，严谨高效，为建立高质量、高水平的专业教材体系奠定了基础。

2）工学结合，围绕学生职业技能设计教材内容和编写形式。基础课程教材在保持扎实理论基础的同时，增加实训、习题、知识拓展以及立体化配套资源；专业课程教材突出理论和实践相统一，注重以企业真实生产项目、典型工作任务、案例等为载体组织教学单元，采用项目导向、任务驱动等编写模式，强调实践性。

3）教材内容科学先进，教材编排展现力强。系列教材紧随技术和经济的发展而更新，及时将新知识、新技术、新工艺和新案例等引入教材；同时注重吸收最新的教学理念，并积极支持新专业的教材建设。教材编排注重图、文、表并茂，生动活泼，形式新颖；名称、名词、术语等均符合国家标准和规范。

4）注重立体化资源建设。系列教材针对部分课程特点，力求通过随书二维码等形式，将教学视频、仿真动画、案例拓展、习题试卷及解答等教学资源融入到教材中，使学生的学习课上课下相结合，为高素质技能型人才的培养提供更多的教学手段。由于我国高等职业教育改革和发展的速度很快，加之我们的水平和经验有限，因此在教材的编写和出版过程中难免出现疏漏。恳请使用本系列教材的师生及时向我们反馈相关信息，以利于我们今后不断提高教材的出版质量，为广大师生提供更多、更适用的教材。

机械工业出版社

前　言

　　随着科技的进步和社会的发展，我国对专业人才英语能力的要求越来越高。电子与通信工程是当今世界发展最迅速、技术更新最活跃的领域之一。我国在该领域注重引进世界先进技术和设备，同时要发展和创造外向型经济，因此，该领域对具有专业英语能力人才的需求更加迫切。为了更好地培养学生的专业外语能力，促进具有国际竞争力的人才培养，编者在追求通俗易懂、简明扼要、便于教学和自学的指导思想下对第 1 版教材进行了改版。

　　本书由 4 部分组成，即电子技术基础、通信技术、计算机技术和高级电子与通信工程。内容基本涵盖了电子技术基础、仪器仪表的使用与维护、计算机技术应用、网络与多媒体技术、通信技术、物联网等领域，同时还收录了一些电子信息技术发展前沿的文章，如人工智能、5G 技术、工业 4.0 与中国制造 2025、3D 打印（增材制造）等。本书内容丰富、题材广泛，语言通俗易懂，能满足高职类不同层次的学生对专业英语的学习需求。

　　本书共分为 22 个单元，每个单元包括课文、生词、专业术语、课文解析、翻译技巧和阅读材料。本书第 17、18 单元，更新了开发学生视野的重要信息：专业网站介绍、英文网站注册申请表填写和中外著名电子信息公司简介；第 3、17 单元选编了两篇创新名人传记的阅读材料，目的是激励和培养高职生的创新能力。最后 4 个单元用相当篇幅分别介绍了英语招聘广告的阅读、个人简历和求职信的书写、面试技巧等，目的是为了使毕业生在就业竞争中能胜人一筹。附录中汇编了十几篇职业现场的交际对话、各单元的参考译文与部分习题答案、常用专业术语缩略语等内容。

　　本书可作为高职高专院校电子类、通信类和计算机控制等专业学生的英语教材，也可供相关专业的工程技术人员参考学习。每单元参考学时为 2~3 学时。建议教师根据学生的接受能力和本校学时情况选用本书 15~20 个单元的内容。附录中的交际英语对话内容，教师可布置给学生主要在课后完成，但教师应有计划地抽查，并占用少量课堂时间做好演示。同时本书应采用生动活泼、灵活多样的互动式教学与课后练习讨论，可多方位培养学生的专业英语学习兴趣与应用能力。

　　本书由河南工业职业技术学院徐存善、王存才、席东河任主编，周志宇、唐红莲、段峰松任副主编，蒋志豪、谭胡心、徐起参编。编写分工为：席东河（河南工业职业技术学院）编写第 1~4 单元和附录 D；段峰松（河南工业职业技术学院）编写第 5~6 单元；王存才（河南工业职业技术学院）编写第 7~9 单元；徐起（浙江省德清尚羽绣纺织品有限公司）编写第 10~11 单元；唐红莲（漯河职业技术学院）编写第 12~14 单元；谭胡心（平顶山工业职业技术学院）编写第 15~17 单元；徐存善（河南工业职业技术学院）编写第 18~19 单元和附录 C；周志宇（廊坊东方职业技术学院）编写第 20~22 单元；蒋志豪（河南工业职业技术学院）编写附录 A；附录 B 中各单元参考译文和习题答案分别由对应的编者编写。

　　本书的编写工作得到了编者所在院校领导的高度重视与大力支持，他们为本书的编写提出了许多宝贵的意见，在此表示衷心的感谢。

　　由于编者水平有限，加之时间仓促，书中难免有不当之处，恳请广大读者和同行批评指正。

编　者

目 录

出版说明
前　言

Chapter Ⅰ　Fundamentals of Electronics Technology

Unit 1　Current, Voltage and Resistance ... 1
　Translating Skills　科技英语翻译的标准与方法 ... 4
　Reading　Conductors, Insulators and Semiconductors ... 6
Unit 2　The Electronic Components ... 8
　Translating Skills　词义的确定 ... 12
　Reading　Basic Resistor Circuits ... 13
Unit 3　The Transistor and Its Basic Circuit ... 16
　Translating Skills　引申译法 ... 19
　Reading for Celebrity Biography（Ⅰ）　Nikola Tesla ... 19
Unit 4　Integrated Circuit ... 22
　Translating Skills　词性转换 ... 24
　Reading　Digital Circuit ... 25
Unit 5　The Electronic Instruments ... 28
　Translating Skills　增词译法 ... 31
　Reading　How to Use a Tester ... 32
Unit 6　Portable Media Player ... 35
　Translating Skills　减词译法 ... 38
　Reading　Digital Audio ... 38

Chapter Ⅱ　Communication Technology

Unit 7　Optical Fiber Communications ... 42
　Translating Skills　科技英语词汇的结构特征（Ⅰ） ... 45
　Reading　Wireless Communications ... 46
Unit 8　Satellite Communications ... 49
　Translating Skills　科技英语词汇的结构特征（Ⅱ） ... 52
　Reading　Internet Telephony & VoIP（Voice over Internet Protocol） ... 53
Unit 9　Global Positioning System（GPS） ... 55
　Translating Skills　被动语态的译法 ... 59
　Reading　Videoconferencing ... 61
Unit 10　4G Network Technology ... 63
　Translating Skills　非谓语动词 V-ing 的用法 ... 66
　Reading　I-Mode ... 67
Unit 11　Smartphone ... 70
　Translating Skills　非谓语动词 V-ed 和 to V 的用法 ... 73

Reading　E-mail & Instant Messaging ·· 74
Unit 12　5G Technology ·· 77
　　Translating Skills　定语从句的翻译 ··· 79
　　Reading　5G Cellular Systems Overview ··· 81

Chapter Ⅲ　Computer Technology

Unit 13　Computer Systems ·· 83
　　Translating Skills　And 引导的句型的译法 ··· 86
　　Reading　BIOS（Basic Input/Output System） ·· 87
Unit 14　Computer Operating System ··· 89
　　Translating Skills　虚拟语气的翻译 ··· 91
　　Reading　Windows XP ·· 92
Unit 15　Programming Language ··· 95
　　Translating Skills　长难句的翻译 ··· 97
　　Reading　MATLAB Language ·· 99
Unit 16　Multimedia Technology ·· 102
　　Translating Skills　电子产品的英文说明书 ··· 106
　　Reading　MIDI Interface ·· 108
Unit 17　China's Progress in Supercomputing ·· 110
　　Useful Information　专业网站介绍与英文网站注册申请表的填写 ··· 113
　　Reading for Celebrity Biography（Ⅱ）　Bill Gates ··· 116

Chapter Ⅳ　Advanced Electronic & Communicative Engineering

Unit 18　Artificial Intelligence（AI） ··· 118
　　Useful Information　中外著名电子信息公司简介 ··· 121
　　Reading　Computer Vision ··· 122
Unit 19　Sensor Technology ·· 125
　　Practical English　怎样阅读英文招聘广告 ·· 129
　　Reading　Remote Sensing ·· 131
Unit 20　Internet of Things ··· 134
　　Practical English　怎样用英文写个人简历 ·· 138
　　Reading　Bluetooth Technology ·· 140
Unit 21　Industry 4.0 Introduction ·· 143
　　Practical English　怎样写英文求职信 ··· 146
　　Reading　*Made in China 2025* and Industrie 4.0 Cooperative Opportunities ···················· 148
Unit 22　3D Printing ·· 151
　　Practical English　面试技巧 ··· 155
　　Reading　American Scientists Work on Printing of Living Tissue Replacements ············· 157
Appendix ··· 160
　　Appendix A　Communication Skills Training for Careers ··· 160
　　Appendix B　Translation and Keys to the Part of Exercises ·· 169
　　Appendix C　New Words List ··· 202
　　Appendix D　Widely Used Abbreviations for Technical Terms ··· 220
参考文献 ··· 225

Ⅶ

Chapter I Fundamentals of Electronics Technology

Unit 1 Current, Voltage and Resistance

Text

The primary purpose of an electric circuit is to move or transfer charges along specified paths.[1] This motion of charges constitutes an electric current. Current flow is represented by the letter symbol I, and it is the time rate of change of charge, given by $I = dq/dt$. The basic unit in which current is measured is ampere (A). An ampere of current is defined as the movement of one coulomb (6.28×10^{18} electrons) past any point of a conductor during one second of time.[2] The milliampere (mA) and the microampere (μA) units are also used to express a magnitude of current much smaller than the ampere. One milliampere is equivalent to one-thousandth of an ampere, and one microampere is equivalent to one – millionth of an ampere.

We shall define voltage "across" an element as the work done in moving a unit charge (+1 C) through the element from one terminal to the other. The term voltage (represented by the letter symbol U) is commonly used to indicate both a potential difference and an electromotive force. The unit in which voltage is measured is volt (V).[3] One volt is defined as that magnitude of electromotive force required cause a current of one ampere to pass through a conductor having a resistance of one ohm.[4] Besides the volt, smaller or larger magnitude of voltage are expressed in millivolt (mV), microvolt (μV) or kilovolt (kV).

Resistors restrict the flow of electric current, for example, a resistor is placed in series with a light-emitting diode (LED) to limit the current passing through the LED.[5] The value of resistor is called resistance and is represented by the letter symbol R. Resistance is measured in ohms; the symbol of ohm is omega (Ω). One ohm is defined as that amount of resistance that will limit the current in a conductor is one ampere when the voltage applied to the conductor is one volt.[6] 1 Ω is quite small, so resistor values are often given in kiloohm (kΩ) or megohm (MΩ).

"Ohm's Law" is one of the fundamental laws of electronics, and pertains to the relationship between current, voltage and resistance in an electrical conductor. This relationship states that "Current = Voltage/Resistance". Ohm's Law states that the ratio of the voltage between the ends of a wire and the current flowing in it is equal to the resistance of the wire. The usual way of expressing this in mathematical terms is "$I = U/R$", and "$U = IR$" or "$R = U/I$" are also used.

New Words and Phrases

current ['kʌrənt] n. (水、气、电) 流；趋势 adj. 现在的；通用的
constitute ['kɔnstitjuːt] vt. 组成，构成；任命；建立
represent [ˌrepri'zent] vt. 表现；代表；表示 vi. 提出异议；代表

voltage   [ˈvəultidʒ]   n. 电压
diode   [ˈdaiəud]   n. 二极管
circuit   [ˈsəːkit]   n. 电路
transfer   [trænsˈfəː]   vt. 转移，移动
charge   [tʃaːdʒ]   n. 费用；负载；电荷   vt. 使充电   vi. 索价；充电
specify   [ˈspesifai]   vt. 指定；列举；详细说明；把……列入说明书
ampere   [ˈæmpɛə]   n. 安［培］
milliampere   [ˈmiliˈæmpɛə]   n. 毫安
microampere   [ˈmaikrəuˈæmpɛə]   n. 微安
coulomb   [ˈkuːlɔm]   n. 库［仑］
magnitude   [ˈmægnitjuːd]   n. 数量，大小，幅度
element   [ˈelimənt]   n. 元件，元素
terminal   [ˈtəːminl]   n. 终端；接线端
ohm   [əum]   n. 欧［姆］
megohm   [ˈmegəum]   n. 兆欧
millivolt   [ˈmilivəult]   n. 毫伏
microvolt   [ˈmaikrəuvəult]   n. 微伏
kilovolt   [ˈkiləuvəult]   n. 千伏
pertain   [pəːˈtein, pə-]   vi. 适合；属于；关于
resistor   [riˈzistə]   n. 电阻器
ratio   [ˈreiʃiəu, -ʃəu]   n. 比，比率
be represented by   用……表示
be equivalent to   等于
in series with   和……串联
pertain to   适合于

Technical Terms

electric circuit   电子电路
potential difference   电位差
electromotive force   电动势
light-emitting diode (LED)   发光二极管
Ohm's Law   欧姆定律

Notes

[1] The primary purpose of an electric circuit is to move or transfer charges along specified paths.

译文：电路的主要作用是沿着特定路径移动或传送电荷。

说明：句中两个表示选择关系的不定式短语作表语。

transfer 除用作动词外，还可以用作名词，如 data transfer 数据传送/转换；file transfer 文件传输；档案传输；文件传送/转移。

They will be offered transfers to other locations. 他们将得到去其他地方的调动令。

[2] An ampere of current is defined as the movement of one coulomb (6.28 × 10^{18} electrons) past any point of a conductor during one second of time.

译文：1A 电流的定义是：在 1s 内 1C（6.28×10^{18}个电子）电量通过导体的任何一点时的电流。

[3] The unit in which voltage is measured is volt (V).

译文：度量电压的单位是伏［特］（V）。

说明：句中 in which...measured 是介词+关系代词引导的定语从句，修饰 unit。

[4] One volt is defined as that magnitude of electromotive force required cause a current of one ampere to pass through a conductor having a resistance of one ohm.

译文：使 1A 电流流过电阻为 1Ω 的导体所需的电动势被定义为 1V。

说明：句中 required 充当后置定语，修饰 electromotive force；having a resistance of one ohm 是现在分词短语作后置定语，修饰 conductor.

magnitude 意思是"量，量值"，如：DC magnitude 直流幅度。

We actually do use the duration of shakingto estimate the magnitude for some small earthquakes. 我们的确会使用震动持续时间来估计一些小型地震的震级。

[5] For example, a resistor is placed in series with a light-emitting diode (LED) to limit the current passing through the LED.

译文：比如，与发光二极管（LED）串联的电阻器限制了流过 LED 的电流。

说明：句中，in series 是介词短语作方式状语。

[6] One ohm is defined as that amount of resistance that will limit the current in a conductor is one ampere when the voltage applied to the conductor is one volt.

译文：1Ω 的定义是：1V 的电压施加在导体上产生 1A 的电流所对应的导体的电阻值。

说明：句中，第二个 that 引导定语从句，修饰 amount of resistance；when 引导时间状语从句；be defined as 表示"被定义为"，如：

Inventory can be defined as "stocks used to support production, support activities and customer service". 库存定义为支持生产、经营以及客户服务的存储。

Exercises

Ⅰ. **Answer the following questions according to the text.**

1. What forms an electric current?
2. Which symbol is used to indicate both a potential difference and an electromotive force?
3. Besides the volt, what are expressed in millivolt (mV), microvolt (μV) or kilovolt (kV)?
4. How is a resistor placed with a light-emitting diode (LED) to restrict the current passing through the LED?
5. Can you describe the Ohm's Law? What is it?

Ⅱ. **Decide whether the following statements are true (T) or false (F) according to the text.**

1. The flow of electrons through a conductor is called an electric current.
2. The term voltage is usually used to show a potential difference but not electromotive force.
3. One millivolt is equivalent to one-thousandth (0.001) of a volt, and one microvolt is equivalent to one-billionth (0.000000001) of a volt.
4. 1 Ω is quite small so resistor values are often given in kiloohm (kΩ) or microohm (μΩ).
5. Ohm's Law expresses that the ratio of the voltage between the ends of a wire and the current flowing in it equals the resistance of the wire.
6. The unit in which resistance is measured is ohm (Ω).

Ⅲ. **Choose the best answer for the following sentences.**

1. Electrons, as one knows, are minute（微小的）_____ charges of electricity.
   A. reverse   B. positive   C. negative
2. The unit of voltage, or potential difference, as _____ is sometimes called, is the volt.
   A. who   B. they   C. it
3. The switch, resistor（电阻器）and wire _____ a circuit.
   A. made of   B. are composed of   C. constitute
4. Matter is made up of atoms, which _____ a number of fundamental particles.
   A. are composed of   B. comprised   C. compose of
5. The flow of electrons _____ electric current.
   A. is made up of   B. makes up   C. consist of
6. A resistor is an electrical component that _____ the flow of electrical current.
   A. increases   B. resists   C. changes
7. The electromotive force or EMF is measured _____ volts.
   A. in   B. on   C. at
8. "Ohm's Law" _____ the relationship between current, voltage and resistance in an electrical conductor.
   A. pertains to   B. state   C. described

Ⅳ. **Translate the following sentences into English.**

1. 电压这个术语常用来表示电位差和电动势。
2. 电阻器的大小叫作电阻。
3. 沿指定路径移动或传送电荷就形成电流。
4. 欧姆定律是电子学中最基本的定律之一。

Translating Skills

科技英语翻译的标准与方法

翻译是再创造，即译者根据原作者的思想，用另一种语言表达出原作者的意思。这就要求译者必须确切理解和掌握原作的内容与含意，在此基础上，很好地运用译文语言把原文的内涵通顺流畅地再现给读者。

1. 翻译的标准

科技英语的翻译标准可概括为"忠实、通顺"4个字。

忠实，首先指忠实于原文内容，译者必须把原作的内容完整而准确地表达出来，不得任意发挥或增删；忠实还指保持原作风格，尽量表现其本来面目。

通顺，即指译文语言必须通俗易懂，符合规范。

忠实与通顺是相辅相成的，缺一不可。忠实而不通顺，读者会看不懂；通顺而不忠实，脱离原作的内容与风格，通顺就失去了意义。例如：

The electric resistance is measured in ohms.

误译：电的反抗是用欧姆测量的。

正译：电阻的测量单位是欧姆。

All metals do not conduct electricity equally well.

误译：全部金属不导电得相等好。

正译：并非所有的金属都同样好地导电。

Some special alloy steels should be used for such parts because the alloying elements make them tougher, stronger, or harder than carbon steels.

误译：对这类零件可采用某些特殊的合金钢，因为合金元素能使它们更加坚韧与坚硬。

正译：对这类零件可采用某些特殊的合金钢，因为合金元素能提高钢的韧性、强度、硬度。

从以上例句可以清楚地看到，不能任意删改，并不等于逐词死译；汉语译文规范化，并非是离开原文随意发挥。此外，还应注意通用术语的译法。比如，第1句中的"The electric resistance"译为"电阻"已成为固定译法，不能用别的译法。

2. 翻译的方法

翻译的方法一般来说有直译（literal translation）和意译（free translation）。直译，即指"既忠实于原文内容，又忠实于原文形式"的翻译；意译，就是指忠实于原文的内容，但不拘泥于原文的形式。

翻译时应灵活运用上述两种方法，能直译的就直译，需要意译的就意译。对同一个句子来说，有时并非只能用一种方法，可以交替使用或同时并用以上两种方法。

请看下面的例子。

Milky Way 应译为"银河"（意译），不可直译为"牛奶路"。

bull's eye 应译为"靶心"（意译），不可直译为"牛眼睛"。

New uses have been found for old metals, and new alloys have been made to satisfy new demands. 老的金属有了新用途，新的金属被冶炼出来，以满足新的需要（本句前半部分用了意译法，后半部分用了直译法）。

The ability to program these devices will make a student an invaluable asset to the growing electronic industry. 编程这些器件的能力将使学生成为日益增长的电子工业领域中的无价人才（这里 asset 原意为"资产"，根据上下文意译成"人才"）。

3. 翻译中的专业性特点

科学技术本身的性质要求科技英语与专业内容相互配合，相互一致，这就决定了专业英语与普通英语有很大的差异。专业英语以其独特的语体，明确表达作者在专业方面的见解，其表达方式直截了当，用词简练。即使同一个词，在不同学科的专业英语中含义也是不同的。例如：

The computer took over an immense range of tasks from workers' muscles and brains.

误译：计算机代替了工人大量的肌肉和大脑。

正译：计算机取代了工人大量的体力和脑力劳动。

（这里 muscles and brains 引申为"体力和脑力劳动"。）

Vibration has worked some connection lose.

误译：振动影响了一些连接的松弛。

正译：振动使一些接线松了。

（connection 可以意译为"接线"或"连接线"。）

In any cases work doesn't include time, but power does.

误译：在任何情况下，工作不包含时间，但功率包含时间。

正译：在任何情况下，功不包括时间，但功率包括时间。

（这里 work、power 在物理专业分别译为"功""功率"。）

Like charges repel each other while opposite charges attracted.

误译：同样的负载相排斥，相反的负载相吸引。
正译：同性电荷相排斥，异性电荷相吸引。
（charge 含义有"负载、充电、充气、电荷"，按专业知识理解为"电荷"。）

从以上例句可知，专业英语专业性强，逻辑性强，翻译要力求准确、精练、正式。这不仅要求我们能熟练地运用汉语表达方式，还要求具有较高的专业水平。

Reading

Conductors, Insulators and Semiconductors

Any substance, which allows electrons to flow freely through its structure, is called a conductor. In general, metals are good conductors. A definite relationship exists between good conductors and their atomic structures. In good conductors, the outer-ring electrons, which are also called valence electrons, may be released from their orbits with relative ease. Atom with 1, 2 and 3 outer ring electrons, and therefore most metals are good conductors.

Substances which prevent the passage of electrons through their structures are called insulators. Insulators have very few easily removed electrons in their outer rings. There are no perfect insulators: first, because of the presence of impurities (foreign materials) which can never be entirely removed; and second, because even a small amount of heat will cause a certain number of valence electrons to be freed from their atoms.

Insulators generally have very stable atomic structures, of which the 4-electron outer-ring structure is typical. In such a structure, there is an absence of easily removed electrons. Examples of good insulators are certain compounds of carbon and diamond, which has a similar atomic structure.

Semiconductors are a group of materials, which conduct electrons poorly and therefore cannot properly be classified either as conductors or insulators. Generally, semiconductors differ from insulators in that their outer-ring electrons can detach themselves from their orbits more easily than in insulators. Typical semiconductors materials are germanium and silicon.

Impurities may be added to pure semiconductors. This results in semiconductor materials, which may either have an excess of free electrons or a deficiency of orbital electrons. When an excess of electrons is present we call the material N-type; when lack of orbital electrons occurs, we call the material P-type. Both N-type and P-type semiconductors are made by treated materials. The addition of impurities to semiconductors is called doping.

New Words and Phrases

insulator ['insjuleitə] *n.* 绝缘体；从事绝缘工作的工人
atomic [ə'tɔmik] *adj.* 原子的，原子能的；微粒子的
valence ['veiləns] *n.* 价；原子价；化合价；效价
impurity [im'pjuərəti] *n.* 杂质；不纯；不洁
diamond ['daiəmənd] *n.* 钻石，金刚石；菱形；*adj.* 金刚钻的
conduct ['kɔndʌkt] *vi.* 带领；导电 *vt.* 管理 *n.* 行为；实施
compound ['kɔmpaund] *vt.* 混合；合成 *n.* 化合物；混合物
detach [di'tætʃ] *vt.* 分离；派遣；使超然
classify ['klæsifai] *vt.* 分类；分等

germanium　　［dʒəːˈmeiniəm］　　n. 锗；锗元素（32号元素，符号Ge）
silicon　　［ˈsilikən］　　n. 硅；硅元素（14号元素，符号Si）
deficiency　　［diˈfiʃənsi］　　n. 缺乏；不足的数额；缺陷，缺点
orbital　　［ˈɔːbitəl］　　adj. 轨道的；眼窝的
excess　　［ikˈses, ˈekses］　　a. 过量的；附加的　　n. 超过，超越，过度，过量
doping　　［ˈdəupiŋ］　　n.（半导体）掺杂质，加添加剂；涂上航空涂料
outer-ring　　外环，外层，外圈，外包围圈
N-type　　N型

Exercises

Ⅰ. **Answer the following questions according to the text.**
  1. What is called a conductor or an insulator?
  2. Why are metals generally good conductors?
  3. Why are there no perfect insulators?
  4. What are typical semiconductor materials?
  5. How to make both N-type and P-type semiconductors? What is called doping?

Ⅱ. **Translate the following phrases and expressions.**
  1. a number of valence electrons
  2. be released from their orbits with relative ease
  3. the 4-electron outer-ring structure
  4. not be classified either as conductors or insulators
  5. either have an excess of free electrons or a deficiency of orbital electrons

Unit 2 The Electronic Components

Text

There are a large number of symbols which represent an equally large range of electronic components. It is important that you can recognize more common components and understand what they actuallydo. [1] A number of these components are drawn below and it is interesting to note that often there is more than one symbol representing the same type of component.

1. Resistors

Resistors restrict the flow of electric current, for example, a resistor is placed in series with a light-emitting diode (LED) to limit the current passing through the LED. Fig. 2-1 shows resistor example and circuit symbol.

Fig. 2-1 Resistor Example and Circuit Symbol

Resistors may be connected in any way round. They are not damaged by heat when soldering. [2]

2. Capacitors

Capacitors store electric charge. They are often used in filter circuits because capacitors easily passAC (changing) signals but they block DC (constant) signals. [3] Fig. 2-2 shows capacitor examples and circuit symbol.

Fig. 2-2 Capacitor Examples and Circuit Symbol

3. Inductor

An inductor is a passive electronic component that stores energy in the form of a magnetic field. An inductor is a coil of wire with many windings, often wound around a core made of a magnetic material, like iron. Fig. 2-3 shows inductor examples and circuit symbol.

Fig. 2-3 Inductor Examples and Circuit Symbol

4. Diodes

Diodes allow electricity to flow in only one direction. The arrow of the circuit symbol shows the direction in which the current can flow. Diodes are the electrical version of a valve and early diodes were actually called valves. Fig. 2-4 shows diode examples and circuit symbol.

Fig. 2-4 Diode Examples and Circuit Symbol

5. Transistors

There are two types of standard transistors, NPN and PNP, with different circuit symbols. The letters refer to the layers of semiconductor material used to make the transistor. Fig. 2-5 shows transistor examples and circuit symbols.

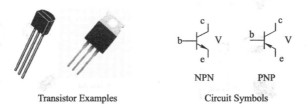

Fig. 2-5 Transistor Examples and Circuit Symbols

6. Integrated Circuits (Chips)

Integrated Circuits are usually called ICs or chips. They are complex circuits which have been etched onto tiny chips of semiconductor (silicon). The chip is packaged in a plastic holder with pins spaced on a 0.1″ (2.54mm) grid which will fit the holes on strip board and breadboards. [4] Very fine wires inside the package link the chip to the pins. Fig. 2-6 shows integrated circuits example.

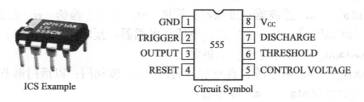

Fig. 2-6 Integrated Circuits Example and Circuit Symbol

7. Light Emitting Diodes (LEDs)

LEDs emit light when an electric current passes through them.

LEDs must be connected in the correct way round, the diagram may be labelled "a" or "+" for anode and "k" or "−" for cathode (yes, it really is "k", not "c", for cathode!). The cathode is the short lead and there may be a slight flat on the body of round LEDs. [5] Fig. 2-7 shows LED examples and circuit symbol.

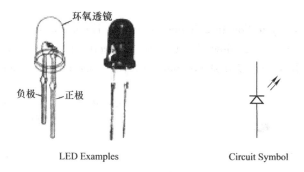

Fig. 2-7　LED Examples and Circuit Symbol

8. Other Electronic Components（Seen in Fig. 2-8）

Fig. 2-8 Shows other electronic components examples and circuit symbols.

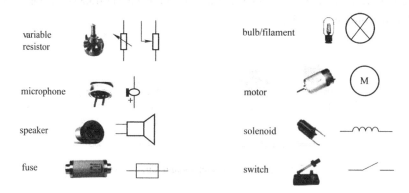

Fig. 2-8　Other Electronic Components Examples and Circuit Symbols

New Words and Phrases

current   ['kʌr(ə)nt]   n. 电流；趋势；涌流 adj. 现在的；通用的；最近的
capacitor   [kə'pæsɪtə]   n. 电容，电容器
filter   ['fɪltə]   n. 滤波器；过滤器；筛选 vt. 过滤，渗透 vi. 滤过，渗入
inductor   [ɪn'dʌktə]   n. 感应器，电感，电感器；授职者；感应体；扼流圈
diode   ['daɪəʊd]   n. [电子] 二极管
valve   [vælv]   n. 电子管，真空管，阀门 vt. 装阀门；以活门调节
transistor [træn'zɪstə]   n. 晶体管
constant   ['kɒnst(ə)nt]   adj. 不变的，恒定的，经常的 n. [数] 常数，恒量
chip   [tʃɪp]   n. 芯片，筹码，碎片；薄片 vt. 削，凿；vi. 剥落，碎裂
semiconductor [ˌsemɪkən'dʌktə] n. 半导体
silicon   ['sɪlɪk(ə)n]   n. 硅，硅元素
winding   ['waɪndɪŋ]   n. 线圈，弯曲，缠绕物 adj. 弯曲的，蜿蜒的；卷绕的
anode   ['ænəʊd]   n. 阳极（电解），正极（原电池）
cathode   ['kæθəʊd]   n. 阴极（电解），负极（原电池）
etched   ['etʃɪd]   adj. 被侵蚀的，风化的 v. 蚀刻（etch 的过去分词）

fuse [fjuːz] n. 熔丝，熔线 vt. 使融合，使熔化 vi. 融合，熔化
filament [ˈfɪləm(ə)nt] n. 灯丝，细线，单纤维；灯丝；细丝
strip [strɪp] n. 带，条状 vt. 剥夺，剥去
solenoid [ˈsəʊlənɒɪd] n. 螺线管，螺线形电导管
represent [reprɪˈzent] v. 表现，代表，体现，作为……的代表；描绘；回忆
breadboard [ˈbredbɔːd] n. 擀面板，案板，电路试验板
diagram [ˈdaɪəɡræm] n. 图表，图解，示意图，线图 vt. 用图表示；图解
light-emitting diode 发光二极管
AC（= Alternating Current） 交流电
DC（= Direct Current） 直流电

Notes

[1] It is important that you can recognize more common components and understand what they actually do.

译文：能够识别更多的普通元件并掌握它们的实际用途是很重要的。

说明：此句是复合句，It 是形式主语，that 引导的主语从句作真正的主语，is important 是系表结构作主句的谓语。但是，主语从句本身也是一个复合句，其主句包含有并列谓语动词 can recognize 与 and understand，而 understand 后面跟有 what 引导的宾语从句。

[2] Resistors may be connected in any way round. They are not damaged by heat when soldering.

译文：电阻可以连接在任一回路中。它们不会因焊接产生的高温而损坏。

说明：这是意义关联的两个简单句。They 指代 Resistors；第 2 句中 when soldering 相当于一个省略了（they are）时间状语从句，当从句主语与主句主语相同，而且助动词是 be 一类的词时，往往省略。

[3] They are often used in filter circuits because capacitors easily pass AC (changing) signals but they block DC (constant) signals.

译文：因为电容器使交流信号容易通过而阻隔直流信号，所以它们经常被用在滤波电路中。

说明：这是一个复合句。because 引导原因状语从句，但从句本身包含了具有并列关系和转折含义的两个分句。

[4] The chip is packaged in a plastic holder with pins spaced on a 0.1″ (2.54mm) grid which will fit the holes on strip board andbreadboards.

译文：该芯片被封装在一个塑料固定物上，引脚间隔距离为 0.1″ (2.54mm)，这样的栅格适合带形板和面包板的孔距。

说明：这是一个复合句。which 引导的定语从句修饰先行词 grid；with pins spaced on a 0.1″ (2.54mm) grid 是介词短语作后置定语修饰 holder。

[5] The cathode is the short lead and there may be a slight flat on the body of round LEDs.

译文：阴极是短的引脚并且在 LEDs 圆形体内可能是微小扁平的那端。

说明：这是一个并列句，两个分句都是简单句。在第二个分句中，on the body of round LEDs 是介词短语作状语。

Exercises

Ⅰ. Write T (True) or F (False) beside the following statements about the text.

1. A resistor is placed in series with a light-emitting diode (LED) to restrict the current pass-

ing through the LED.

   2. Capacitors easily pass DC signals but they block AC signals.
   3. Capacitors are often used in filter circuits.
   4. Integrated Circuits are usually referred to as ICs or chips.
   5. There are two kinds of standard transistors, NPN and PNP.
   6. LEDs send out light when an electric current passes through them.
   7. ICs are complex circuits which have been etched onto tiny chips of semiconductor (silicon).

Ⅱ. Match the following words on the left with the appropriate definition or expression on the right.

   1. resistor   a. a device for controlling the flow of a liquid which letting it move in one direction only
   2. inductor   b. a piece of wire, wound into circles, which acts as a magnet when carrying an electric current
   3. valve   c. a small wire or device inside a piece of electrical equipment that breaks and stops the current if the flow of electricity is too strong
   4. fuse   d. a device that has resistance to an electric current in a circuit
   5. solenoid   e. a component in an electric or electronic circuit which possesses inductance

Ⅲ. Fill in the missing words according to the text.

   1. _____ restrict the flow of electric current.
   2. Capacitors are often used in _____ circuits because they easily pass AC signals but they block DC signals.
   3. Diodes are the electrical _____ of a valve and early diodes were actually called valves.
   4. The chip is packaged in a _____ holder with pins spaced on a 0.1" (2.54mm) grid which will fit the holes.
   5. Diodes allow electricity to flow in _____ direction.

Ⅳ. Translate the following sentences into English.

   1. 能够识别更多的电子元器件并掌握它们的实际用途是非常必要的。
   2. 电阻器不会因焊接高温而损坏。
   3. 金属要加热到一定温度才能熔化。
   4. 电子学领域包括电子管、晶体管和集成电路等。

Translating Skills

词义的确定

英、汉两种语言都有一词多性、一词多义的现象。一词多性是指一个词往往具有多个词性，具有几种不同的意义。例如，display 既可作名词表示"显示（器）"，又可以作形容词表示"展览的、陈列用的"，还可作动词表示"显示、表现"等意思。一词多义是说在同一词性中，往往有几个不同的词义。例如，power 这个词，作名词的意思包括"电力、功率、乘方"等。在翻译过程中，在弄清句子结构后就要善于选择和确定句子中关键词的词性和意义。选择和确定词义通常从以下几个方面入手。

1. 根据词性确定词义

确定某个词的词义时，首先要确定这个词在句中应属于哪一种词性，然后再进一步确定其词义。下面以 display 为例：

Here, you have the option of defining your own display variants. 这里，你有权定义你自己的显示形式。（display 为名词）

Often, it is best to display materials on an information table. 通常，最好是把资料展示在提供各类资讯的桌子上。（display 为动词）

The reverse side of a control panel, display panel, or the like is the side with the interconnecting wiring. 控制面板、显示面板或类似的面板的反面，是带有互连接线的那一面。（display 为形容词）

2. 同一词性表达不同词义

英语中同一个词，即使属于同一词性，在不同的场合中也往往具有不同的含义。此时，必须根据上、下文的联系及整个句子的意思加以判断和翻译。例如，as 这个词作连词时有以下用法：

The volume varies as the temperature increases. 体积随着温度增加而变化。（as 引导时间状语从句）

As heat makes things move, it is a form of energy. 因为热能使物体运动，所以热是能的一种形式。（as 引导原因状语从句）

3. 根据单词搭配情况确定词义

英译汉时，不仅必须根据上、下文的联系理解词义，还需要根据词的搭配情况来理解词义。尤其在科技文献中，由于学科及专业不同，同一个词在不同的专业中具有不同的意义。比如，

The fifth power of two is thirty-two. 2 的 5 次方是 32。（数学）

With the development of electrical engineering, power can be transmitted over long distances. 随着电气工程的发展，电力能输送得非常远。（电学）

Friction can cause a loss of power in every machine. 摩擦能引起每一台机器的功率损耗。（物理学）

4. 根据名词的单复数选择词义

英语中有些名词的单数或复数表达的词义完全不同。例如：

名词	单数词义	复数词义
facility	简易，灵巧	设施，工具
charge	负荷，电荷	费用
spirit	精神	酒精

Although they lost, the team played with tremendous spirit. 他们虽然输了，但却表现出了极其顽强的精神。

Whisky, brandy, gin and rum are all spirits. 威士忌、白兰地、杜松子酒和朗姆酒都是烈酒。

Reading

Basic Resistor Circuits

Circuits consisting of just one battery and one load resistance are very simple to analyze, but they are not often found in practical applications.[1] Usually, we find circuits where more than two components are connected together. There are two basic ways in which resistors can be connected: in series and in parallel. A simple series resistance circuit is shown in Fig. 2-9.

Determining the total resistance for two or more resistors in series is very simple. Total resistance equals the sum of the individual resistances. In this case, $R_T = R_1 + R_2$. This makes common sense; if you think again in terms of water flow, a series of obstructions in a pipe add up to slow the flow more than any one. The resistance of a series combination is always greater than any of the individual resistors.

The other method of connecting resistors is shown in Fig. 2-10, which shows a simple parallel resistance circuit.

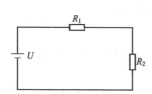 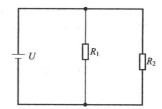

Fig. 2-9 Two Resistors in Series   Fig. 2-10 Two Resistors in Parallel

Our water pipe analogy indicates that it should be easier for current to flow through this multiplicity of paths, even easier than it would be to flow through any single path. [2] Thus, we expect a parallel combination of resistors to have less resistance than any one of the resistors. Some of the total current will flow through R_1 and some will flow through R_2, causing an equal voltage drop across each resistor. More current, however, will flow through the path of least resistance. The formula for total resistance in a parallel circuit is more complex than for a series circuit:

$$\frac{1}{R_T} = \frac{1}{R_1} + \frac{1}{R_2}$$

Parallel and series circuits can be combined to make more complex structures, but the resulting complex resistor circuits can be broken down and analyzed in terms of simple series or parallel circuits. [3] Why would you want to use such combinations? There are several reasons. You might use a combination to get a value of resistance that you needed but did not have in a single resistor. Resistors have a maximum voltage rating, so a series of resistors might be used across a high voltage. Also, several low power resistors can be combined to handle higher power. What type of connection would you use?

And, of course, the complexity doesn't stop at simple series and parallel either! We can have circuits that are a combination of series and parallel, too:

In this circuit (as shown in Fig. 2-11), we have two loops for electrons to flow through: one from 6 to 5 to 2 to 1 and back to 6 again, and another from 6 to 5 to 4 to 3 to 2 to 1 and back to 6 again. Notice how both current paths go through R_1 (from point 2 to point 1). In this configuration, we'd say that R_2 and R_3 are in parallel with each other, while R_1 is in series with the parallel combination of R_2 and R_3.

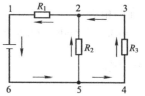

Fig. 2-11 Series-parallel Circuit

New Words and Phrases

analyze [ˈænəlaiz] vt. 分析；分解；解释；对……进行心理分析
obstruction [əbˈstrʌkʃn] n. 障碍物；阻碍物；阻碍；阻挠
combination [kɒmbiˈneiʃn] n. 结合；联合体；密码组合
individual [indiˈvidjuəl] adj. 个人的；独特的；个别的 n. 个人；个体
complex [kɒmbiˈneiʃ(ə)n] adj. 复杂的；难懂的 n. 情结；不正常的忧虑
connection [kəˈnekʃn] n. 连接；联系，关系；连接点
configuration [kənfigəˈreiʃn] n. 布局，构造；配置；排列；［物］位形，组态
multiplicity [mʌltiˈplisiti] n. 多样性

Notes

[1] Circuits consisting of just one battery... are not often found in practical applications.

译文：由一个电池和一个负载电阻组成的电路易于分析，但在实际应用中并不常见。

[2] Our water pipe analogy indicates... than it would be to flow through any single path.

译文：我们用水管来类比并联电路，流体流过多重并联路径比流过任何一个单一路径更容易。

说明：此句是复合句。that 引导的宾语从句本身也是一个复合句，由 even easier than 引出比较状语从句，修饰 should be easier。

[3] Parallel and series circuits... in terms of simple series or parallel circuits.

译文：并联和串联电路可以结合起来，组成更复杂的结构，但由此产生的复杂电阻电路可以根据简单的串联或并联电路进行分解和分析。

Exercises

I. Fill in the blanks with the proper word. Change the form if necessary.

individual	sense	resist	combine
series	in terms of	great	obstruct

Determining the total _____ for two or more resistors in _____ is very simple. Total resistance equals the sum of the _____ resistances. In this case, $r_T = r_1 + r_2$. This makes common _____; if you think again _____ water flow, a series of _____ in a pipe add up to slow the flow more than any one. The resistance of a series _____ is always _____ than any of the individual resistors.

II. Translate the of llowing sentences into English.

1. 操作由特定于服务的消息交换组成。
2. 启动服务器甚至比前面的安装更容易。
3. 串联组合的电阻总是大于任何一个单独的电阻器。

15

Unit 3　The Transistor and Its Basic Circuit

Text

Transistors are the most important device in electronics today. Not only are they made as discrete (separate) components, but also integrated circuits (ICs) may contain thousands of transistors on a tiny slice of silicon.

Transistors are three-terminal devices used as amplifiers and switches. There are two basic types. They are:

1) The junction transistor (usually called the transistor), its operation depends on the flow of both majority and minority carriers and it has two P-N junctions.

2) The field effect transistor (called the FET) in which current is due to majority carriers only (either electrons or holes) and there is just one P-N junction. [1]

A transistor consists of three layers of semiconductor material: a thin layer of one type with the other type on each side. There are two possible arrangements: N-type in the middle with P-type on each side (PNP) and P-type in the middle with N-type on each side (NPN). [2] The center is called base, one outside layer is called the emitter, and the other is known as the collector (as shown in Fig. 3-1). [3]

A transistor is an electronical device that regulates the current flowing through it. Current from a power source enters the emitter, passes through the very thin base region, and leaves via the collector.

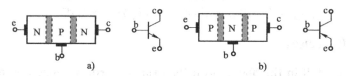

Fig. 3-1　P-N Junction and Transistor

Current flow is always in this direction. This current can be made to vary in amplitude by varying the current flowing in the base circuit. It takes only a small change of base current to control a relatively large collector current. [4] It is this ability that enables the transistor to amplify.

There are three basic ways of connecting transistors in a circuit (as shown in Fig. 3-2): common-base, common-emitter and common-collector. In the common-base connection, the signal is introduced into the emitter-base circuit and extracted from the collector-base circuit. Because the input or emitter-base circuit has a low impedance in the order of 0.5 to 50 ohms and the output or collector base circuit has a high impedance in the order of 1,000 ohms to one megohm, the voltage or power gain in this type of configuration may be in the order of 1,500.

In the common-emitter connection, the signal is introduced into the base-emitter circuit and extracted from the collector-emitter circuit. This configuration has more moderate input and output impedance than the common-base circuit. The input (base-emitter) impedance is in the range of 20 to 5,000 ohms, and output (collector-emitter) impedance is about 50 to 50,000 ohms. Power gain in the order of 10,000 (or about 40dB) can be realized with this circuit because it provides both current gain and voltage gain.

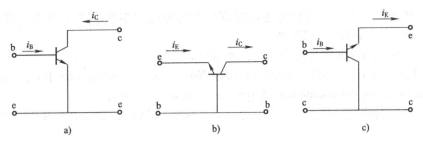

Fig. 3-2 Three Basic ways of Connecting Transistors in a Circuit

The third type of connection is the common-collector circuit. In this configuration, the signal is introduced into the base-collector circuit and extracted from the emitter-collector circuit. Because the input impedance of the transistor is high and the output impedance low in this connection, the voltage gain is less than one and the power gain is usually lower than that obtained in a common-base or a common-emitter circuit.

New Words and Phrases

```
base         [beis]              n. 基极；基础；底部  adj. 低劣的  vt. 以……作基础
emitter      [i'mitə]            n. 发射极，发射体；发射器
collector    [kə'lektə]          n. 集电极；收税员；征收者；收藏家
discrete     [dis'kri:t]         adj. 离散的  n. 分立元件；独立部件
integrate    ['intigreit]        vt. 使……成整体  vi. 成为一体  n. 集成体
slice        [slais]             n. 薄片；菜刀  vt. 切下；将……切成薄片  vi. 切开
amplifier    ['æmplifaiə]        n. 放大器，扩大器；扩音器
junction     ['dʒʌŋkʃən]         n. 连接，接合；交叉点；接合点
minority     [mai'nɔ:rəti]       n. 少数民族；未成年  adj. 属于少数派的
regulate     ['regjuleit]        vt. 调节，规定；有系统地管理；控制；校准
amplitude    ['æmplitju:d]       n. 振幅；广阔；丰富，充足幅度
bias         ['baiəs]            n. 偏见；斜纹；偏置  vt. 使存偏见  adj. 偏斜的
reverse      [ri'və:s]           n. 背面；相反；反向  vt. 颠倒  adj. 反面的
extract      [ik'strækt, 'ekstrækt]  vt. 摘录；输出；取出  n. 摘录
impedance    [im'pi:dəns]        n. 阻抗；全电阻；输入阻抗
gain         [gein]              n. 收获；增益；利润  vt. 获得；增加  vi. 获利；增加
configuration [kən'figju'reiʃən] n. 配置；结构；外形
moderate     ['mɔdəreit]         adj. 适度的；中等的；温和的  v. 节制；减轻
in the order of      约为，大约
in the range of      在……范围内
base-emitter circuit    基 – 射回路
collector-emitter circuit  集 – 射回路
```

Notes

[1] The field effect transistor (called the FET) in which current is due to majority carriers only... and there is just one P-N junction.

译文：场效应晶体管（简称为场效应管）的电流仅由多数载流子提供（可以是电子，也可以是空穴），且只有一个 PN 结。

说明：句中 in which 引导定语从句，修饰 transistor。

[2] There are two possible arrangements: N-type in the middle with P-type on each side (PNP) and P-type in the middle with N-type on each side (NPN).

译文：可以有两种安排：N 型在中间 P 型在两边（PNP），以及 P 型在中间 N 型在两边（NPN）。

说明：N-type... on each side (PNP) 和 P-type... on each side (NPN) 是 arrangements 的同位语。

[3] The center is called base, one outside layer... and the other is known as the collector.

译文：中间称为基极，两边分别称为发射极和集电极。

说明：one... the other... 是一个常见的句型，意思是 "一个……另一个……"。句中 layer 表示 "层，层次" 的意思。例如：

To plate metal is to cover it with a thin layer of gold, or silver, etc. 电镀金属就是把金属薄薄地覆盖上一层金或银等。

[4] It takes only a small change of base current to control a relatively large collector current.

译文：基极电流只需很小的变化，就会引起集电极电流很大的变化。

说明：It takes... to do sth，意思是 "做某事花费某人多少时间（金钱等）"，例如：

It takes a long time to make a film. 拍电影要花很长一段时间。

It takes perseverance to do anything well. 坚持是做好任何事情所需要的一种品质。

Exercises

Ⅰ. Answer the following questions according to the text.

1. What does the operation of the junction transistor depend on? And how many P-N junctions does it have?

2. What is the current of the field effect transistor due to? Does it have two P-N junctions?

3. How does a transistor amplify the current through it?

4. Why can the power gain in the order of 10,000 (or about 40dB) be realized with the common-emitter connection?

Ⅱ. Choose the best answer for the following sentences.

1. Transistors are three-terminal _____ used as amplifiers and as switches.

A. components   B. devices   C. elements

2. Its operation depends on the flow of both _____ and minority carriers and it has two P-N junctions.

A. major   B. minor   C. majority

3. A transistor consists of three _____ of semiconductor material.

A. kinds   B. layers   C. sorts

4. The center is called base, one outside layer is called the emitter, and the other is known as the _____.

A. collecting   B. junction   C. collector

5. In the common-base connection, the signal is introduced into _____ and extracted from the collector-base circuit.

A. emitter-base circuit   B. emitter   C. collector

6. The common-emitter connection has more moderate input & output _____ than the common-base circuit.

A. resistance   B. capacitance   C. impedance

Ⅲ. **Translate the following sentences into English.**

1. 晶体管是电子学中最重要的器件。
2. 晶体管有两种基本类型，即结型晶体管（通常称为晶体管）和场效应晶体管（简称为场效应管）。
3. 晶体管是调节流过它的电流的电子控制器件。
4. 因为晶体管的输入阻抗高而输出阻抗低，所以电路的电压增益比1小，而功率增益比共发射极或共基极连接的要低。

Translating Skills

引 申 译 法

英、汉两种语言在表达上有很大差别。翻译时，有些词或词组不能直接搬用词典中的释义，若生搬硬套，会使译文生硬晦涩，难以看懂，甚至造成误解。所以，要在弄清原文内涵的基础上，根据上下文的逻辑关系和汉语的搭配习惯，对词义加以引申。若遇到专业方面的内容，必须选用专业术语。引申后的词义能更确切地表达原文意义。例如：

However, colors can give more force to the form of the product.

欠佳译法：然而，色彩能给予产品更多的力量。

引申译法：然而，色彩能使产品外形增添美感。

Power plugs are male electrical connectors that fit into female electrical sockets.

欠佳译法：电源插头是雄性连接器适配雌性连接器。

引申译法：电源插头可以插入电源插座。

High-speed grinding does not know this disadvantage.

欠佳译法：高速磨床不知道这个缺点。

引申译法：高速磨床不存在这个缺点。

The charge current depends upon the technology and capacity of the battery being charged.

欠佳译法：充电电流的大小取决于充电技术和被充电电池的容量。

引申译法：充电电流的大小根据充电技术和电池容量的不同而不同。

Reading for Celebrity Biography (Ⅰ)

Nikola Tesla

Nikola Tesla was born in 1856 in Smiljan Lika, Croatia. He was the son of a Serbian Orthodox clergyman. Tesla studied engineering at the Austrian Polytechnic School. He worked as an electrical engineer in Budapest and later emigrated to the United States in 1884 to work at the Edison Machine Works. He died in New York City on January 7, 1943.

During his lifetime, Tesla invented fluorescent lighting, the Tesla induction motor, theTesla coil, and developed the alternating current (AC) electrical supply system that included a motor and transformer, and 3-phase electricity.

Tesla is now credited with inventing modern radio as well; since the Supreme Court overturned Guglielmo Marconi's patent in 1943 in favor of Nikola Tesla's earlier patents. When an engineer (Otis Pond) once said to Tesla, "Looks as if Marconi got the jump on you" regarding Marconi's radio system, Tesla replied, "Marconi is a good fellow. Let him continue. He is using seventeen of my patents."

The Tesla coil, invented in 1891, is still used in radio and television sets and other electronic equipment.

Nikola Tesla —Mystery Invention

Ten years after patenting a successful method for producing alternating current, Nikola Tesla claimed the invention of an electrical generator that would not consume any fuel. [1] This invention has been lost to the public. Tesla stated about his invention that he had harnessed the cosmic rays and caused them to operate a motive device.

In total, Nikola Tesla was granted more than one hundred patents and invented countless unpatented inventions.

Nikola Tesla and George Westinghouse

In 1885, George Westinghouse, head of the Westinghouse Electric Company, bought the patent rights to Tesla's system of dynamos, transformers and motors. Westinghouse used Tesla's alternating current system to light the World's Columbian Exposition of 1893 in Chicago.

Nikola Tesla and Thomas Edison

Nikola Tesla was Thomas Edison's rival at the end of the 19th century. In fact, he was more famous than Edison throughout the 1890s. His invention of polyphase electric power earned him worldwide fame and fortune. At his zenith he was an intimate of poets and scientists, industrialists and financiers. Yet Tesla died destitute, having lost both his fortune and scientific reputation. [2] During his fall from notoriety to obscurity, Tesla created a legacy of genuine invention and prophecy that still fascinates today.

New Words and Expressions

polytechnic　[ˌpɒlɪˈteknɪk]　adj. 综合技术的　n. 理工专科学校
emigrate　[ˈemɪɡreɪt]　vi. 移居；移居外国　vt. 移民
fluorescent　[fluəˈres(ə)nt]　adj. 荧光的；发亮的　n. 荧光灯
overturn　[əʊvəˈtɜːn]　vt. 推翻；倾覆；破坏
grant　[ɡrɑːnt]　vt. 授予；允许；承认　vi. 同意
dynamo　[ˈdaɪnəməʊ]　n. 发电机；精力充沛的人
transformer　[trænsˈfɔːmə]　n. [电] 变压器
polyphase　[ˈpɒlɪfeɪz]　adj. 多相的
zenith　[ˈzenɪθ]　n. 顶峰；顶点；最高点
intimate　[ˈɪntɪmət]　adj. 亲密的；私人的；精通的；n. 知己；至交
destitute　[ˈdestɪtjuːt]　adj. 穷困的；缺乏的　n. 赤贫者　vt. 使穷困；夺去
notoriety　[nəʊtəˈraɪɪtɪ]　n. 声名狼藉；名声远扬；著名人物
obscurity　[əbˈskjʊərɪtɪ]　n. 朦胧；阴暗；晦涩；身份低微；不分明
prophecy　[ˈprɒfɪsɪ]　n. 预言；预言书；预言能力
coupling　[ˈkʌplɪŋ]　n. [电] 耦合；结合，联结　v. 连接

legacy　　　[ˈlegəsɪ]　　n. 遗赠，遗产
cosmic　　　[ˈkɒzmɪk]　　adj. 宇宙的（等于 cosmical）
alternating current　多相电
cosmic ray　宇宙射线

Notes

[1] Ten years after patenting a successful... that would not consume any fuel.

译文：在发明了产生交流电的方法并申请专利 10 年后，尼古拉·特斯拉声称这是发明了一种不需要任何燃料的发电机。

说明：这是包含有定语从句的复合句。主句是 Nikola Tesla claimed ... generator；that 引导定语从句修饰 generator；Ten years... current 是时间状语，修饰 claimed；其中 for producing... 是介词短语作后置定语修饰 method。

[2] Yet Tesla died destitute, having lost both his fortune and scientific reputation.

译文：然而特斯拉去世时贫困潦倒，丢失了财富和在科学界的声誉。

说明：这是一个简单句。died 是不及物动词；destitute 是形容词作方式状语；句中 having lost... 表示伴随情况的结果状语。

Exercises

Ⅰ. Decide whether the following statements are True (T) or False (F) according to the text.

1. Nikola Tesla is the inventor of polyphase alternating current.

2. Tesla was very rich when he died at the year of 1943.

3. Tesla once worked in Edison Machine Works and he wasThomas Edison's rival at the end of the 19th century.

4. Tesla made friends with many successful man at his zenith.

Ⅱ. Translate the following sentences into English.

1. 特斯拉声称他每天从凌晨 3 点工作到晚上 11 点，星期天和节假日也不例外。

2. 1890 年后特斯拉试验用电感和电容与高压交流电耦合的方式传输他的特斯拉线圈生成的电力。

3. 尝试着发明一个生成交流电的更好的方法，特斯拉发明了一个蒸汽驱动的往复式电力发电机并在 1893 年申请专利，同年在哥伦布纪念博览会上开始使用。

Unit 4　Integrated Circuit

Text

An integrated circuit (IC) is a combination of a few interconnected circuit elements such as transistors, diodes, capacitors and resistors. It is a small electronic device made out of a semiconductor material. The first integrated circuit was developed in the 1950s by Jack Kilby of Texas Instruments and Robert Noyce of Fairchild Semiconductor.

The electrically interconnected components that make up an IC are called integrated elements.[1] If an integrated circuit includes only one type of components, it is said to be an assembly or set of components.

Integrated circuits (seen in Fig. 4-1) are used for a variety of devices, including microprocessors, audio and video equipments, and automobiles. Integrated circuits are often classified by the number of transistors and other electronic components, they contain:

- SSI (small-scale integration): Up to 100 electronic components per chip;
- MSI (medium-scale integration): From 100 to 3000 electronic components per chip;
- LSI (large-scale integration): From 3000 to 100000 electronic components per chip;
- VLSI (very large-scale integration): From 100000 to 1000000 electronic components per chip;
- ULSI (ultra large-scale integration): More than 1 million electronic components per chip.

As the capability to integrate a greater number of transistors in a single integrated circuit (IC) grows, it is becoming more common that an application-specific IC (ASIC) is required, at least for high volume applications.[2] Advances in silicon technology have allowed IC designers to integrate more than a few million transistors on a chip; even a whole system of moderate complexity can now be implemented on a single chip.

Fig. 4-1　Integrated Circuits

The invention of IC is a great revolution in the electronic industry. Sharp size, weight reductions are possible with these techniques, and more importantly, high reliability, excellent functional performance, low cost and low power dissipation can be achieved. ICs are widely used in the electronic industry.

New Words and Expressions

　　assembly　［əˈsembli］　n. 装配；集会，集合
　　combination　［ˌkɔmbiˈneiʃən］　n. 结合，联合，合并
　　interconnect　［ˌintəkəˈnektid］　vt. 使互相连接　vi. 互相联系
　　semiconductor　［ˌsemikənˈdʌktə(r)］　n. ［电子］［物］半导体；半导体器件
　　microprocessor　［ˈmaikrəuˈprəusesə(r)］　n. ［计］微处理器
　　equipment　［iˈkwipmənt］　n. 设备，装备；器材，配件

automobile　[ˈɔːtəməbiːl]　n. 〈美〉汽车 vt. 驾驶汽车
contain　[kənˈtein]　vt. 包含，容纳；克制，遏制；牵制
classify　[ˈklæsifaɪ]　vt. 分类，归类；把……列为密件
video　[ˈvidiəʊ]　n. [电子] 视频；录像，录像机　adj. 视频的；录像的
implement　[ˈimplimənt]　vt. 实施，执行；实现，使生效　n. 工具；手段
ultra　[ˈʌltrə]　adj. 极端的；过分的　n. 极端主义者；激进论者
silicon　[ˈsilik(ə)n]　n. [化学] 硅；硅元素
moderate　[ˈmɒdərət]　adj. 有节制的；温和的；适度的；中等的
complexity　[kəmˈpleksəti]　n. 复杂性，复杂的事物；复合物
revolution　[ˌrevəˈluːʃn]　n. 革命；彻底改变；旋转；运行，公转
industry　[ˈindəstri]　n. 工业；产业（经济词汇）；工业界
reliability　[riˌlaiəˈbiləti]　n. 信度；可靠性，可靠度
dissipation　[ˌdisiˈpeiʃn]　n. （物质、精力逐渐的）消散，分散功耗
functional　[ˈfʌŋkʃənl]　adj. 功能的
the circuit element　电路元件
the video equipment　视频设备
electronic component　电子元件
ultra large-scale integration（ULSI）　超大规模集成电路
the electronic industry　电子工业

Notes

[1] The electrically interconnected components that make up an IC are called integrated elements.

译文：电学上把组成集成电路的彼此相连接的元器件称为集成元器件。

说明：句中 that make up an IC 为定语从句，修饰先行词 interconnected components。

[2] As the capability to integrate a greater number of transistors in a single integrated circuit (IC) grows, it is becoming more common that an application-specific IC (ASIC) is required, at least for high volume applications.

译文：随着在一个芯片上集成大量晶体管的能力（即集成电路的集成量）的提高，对专用集成电路的需求已更加普遍。至少对大批量的应用来说更需要专用的集成电路。

说明：这是一个复合句，as 引导一个时间状语从句；主句本身也是一个复合句，it 是形式主语，真正的主语是由 that 引导的主语从句来担任的；at least 引导让步状语。

Exercises

Ⅰ. Answer the following questions.
1. What is an integrated circuit? When was the first integrated circuit developed?
2. Whatare called integrated elements? What is named an assembly or set of components?
3. How many types are integrated circuits often classified? What are they?
4. What are possibly achieved with the invention of IC?

Ⅱ. Decide whether the following statements are True (T) or False (F) according to the text.
1. An integrated circuit is a small electronic device made out of a semiconductor material.
2. A whole system of moderate complexity can now be implemented on a single chip.

3. Integrated circuits are used for a few of devices.
4. Integrated circuits are often classified by the number of transistors and other electronic components.
5. VLSI: up to from 3000 to 10,000 electronic components per chip.
6. The invention of IC is not a great revolution in the electronic industry.

Ⅲ. **Translate the following sentences into English.**
1. 集成电路是几种相互连接的电路元器件在一块半导体芯片上的组合。
2. 把组成集成电路的彼此电气连接的元器件称为集成元器件。
3. 集成电路已广泛应用在电子工业。
4. 集成电路通常根据其包含的晶体管和其他电路元器件的数量来归类。

Ⅳ. **Translate the following short passage into Chinese.**
An integrated circuit looks like nothing more than a tiny chip of metal, perhaps one-half of a centimeter on a side, and not much thicker than a sheet of paper. It is so small that if it fell on the floor, it could be easily swept up with the dust. Although it is very small, it represents the most highly skilled technology at every step of its manufacture. At today's level of development, it might comprise more than ten thousand even several millions of separate electronic elements including elements of many different functions, such as diodes, transistors, capacitors and resistors.

Translating Skills

词 性 转 换

在翻译过程中，由于英、汉两种语言的表达方式不同，不能逐词对译。原文中有些词在译文中须转换词性，才能使译文通顺流畅。词性转换包括以下几种情况。

1. 英语动词、形容词、副词译成汉语名词

Telecommunications <u>means</u> so much in modern life that without it our modern life would be impossible. 电信在现代生活中<u>意义</u>重大，没有它就不可能有我们现在的生活。

The cutting tools must be <u>strong, tough, hard</u> and wear resistant. 刀具必须有足够的<u>强度、韧性、硬度</u>，而且耐磨。

Dynamics is divided into statics and kinetics, the <u>former</u> treating of forces in equilibrium, the <u>latter</u> of the relation of force to motion. 力学分为静力学和动力学：<u>前者</u>研究平衡力，<u>后者</u>研究力和运动的关系。

The image must be <u>dimensionally</u> correct. 图形的<u>尺寸</u>必须正确。

2. 英语名词、介词、形容词、副词译成汉语动词

<u>Substitution</u> of manual finishing is one example of HSM application. <u>替代</u>手工精加工是高速加工应用的一个例子。

Scientists are <u>confident</u> that all matter is indestructible. 科学家们<u>深信</u>一切物质是不灭的。

In any machine input work equals output work <u>plus</u> work done <u>against</u> friction. 任何机器的输入功，都等于输出功<u>加上克服</u>摩擦所做的功。

Open the valve to let air <u>in</u>. 打开阀门，让空气<u>进入</u>。

3. 英语的名词、副词和动词译成汉语形容词

This wave guide tube is <u>chiefly</u> characterized by its simplicity of structure. 这种波导管的<u>主要</u>特点是结构简单。

They said that such knowledge is <u>needed</u> before they can develop a successful early warning system for earthquakes. 他们说，这种知识对他们发明一种有效的地震早期警报<u>是必要的</u>。

4. 英语形容词、名词译成汉语副词

With <u>slight</u> repairs the television transmitters can be used. 只要<u>稍加</u>修理，这台视频发射机就可使用。

A <u>continuous</u> increase in the temperature of a gas confined in a container will lead to a <u>continuous</u> increase in the internal pressure within the gas. <u>不断</u>提高密封容器内气体的温度，会使气体的内压力<u>不断</u>增大。

Reading

Digital Circuit

The phrase "digital electronics" is used to describe those circuit systems which primarily operate with the use of only two different voltage levels or two other binary states.[1] The two different states by which digital circuits operate may be of several forms. They can, in the most simple form, consist of the opening and closing of a switch. In this case, the closed-switch state can be represented by 1 and the open-switch state by 0.[2]

A very common method of digital operation is achieved by using voltage pulses. The presence of a positive pulse can be represented by 1 and the absence of a pulse by 0. With a square-wave signal,[3] the positive pulses can represent 1 and the negative pulses can represent 0.

Integrated circuits containing many transistors are most commonly used as switching devices in digital electronics logic gates. The three basic types of digital logic gates are the AND gate, the OR gate, and the NOT gate (as shown in Fig. 4-2). The operation of an AND gate is mathematically expressed by the equation $A \cdot B = C$. This can be read as "input A and input B equals output C". The operation of an OR gate is often expressed by the equation $A + B = C$. This can be read as "either input A or input B (or both) equals output C". An important function of NOT gate is to produce signal inversion or an output signal that is opposite in nature to the input signal.[4] Any logic function can be performed by the three basic gates that have been described. Even in a large scale digital system, such as a computer, control or digital-communication system, there are only a few basic operations, which must be performed.[5] The three basic types of digital logic gates and the flip-flop are the four circuits most commonly employed in such systems.

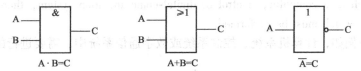

Fig. 4-2 Digital Logic Gates

New Words and Phrases

primarily ['praimərəli] *adv.* 主要地，根本上；首先；最初的
binary ['bainəri] *adj.* 二元的，二态的；二进制的
pulse [pʌls] *n.* 脉搏；脉冲 *vt.* 使跳动 *vi.* 跳动，脉动

represent　　［ˌrepriˈzent］　*vt. vi.* 表现；描绘；代表；回忆；再赠送
nature　　［ˈneitʃə］　*n.* 自然；性质；种类；本性
inversion　　［inˈvəːʃən］　*n.* 倒置；倒转；反向反转
scale　　［skeil］　*n.* 刻度，衡量，数值范围　*v.* 依比例决定；攀登
flip-flop　　［ˈflipflɔp］　*n.* 触发器；啪嗒啪嗒的响声　*vt.* 使翻转；使突然转变
the closed-switch state　开关闭合状态
the open-switch state　开关断开状态
logic gates　［计算机］逻辑闸；逻辑门电路库；逻辑门
AND gate　和门；与门；与电路
binary arithmetic　二进制算术

Notes

［1］The phrase "digital electronics" is used to describe those circuit systems which primarily operate with the use of only two different voltage levels or two other binary states.

译文："数字电子技术"这一术语是用来描述只用两种不同电压电平，或两种不同的二进制状态，进行工作的电路系统。

说明：句中 to describe... or two other binary states 是不定式短语，充当目的状语。which 引导的定语从句修饰 systems。

［2］In this case, the closed-switch state can be represented by 1 and the open-switch state by 0.

译文：在这种情况下，开关闭合状态用1表示，开关断开状态用0表示。

说明：句中 closed-switch 和 open-switch 具有形容词性，充当定语，均修饰 state。and the open-switch state by 0 已省略了谓语动词部分 can be represented。

［3］With a square-wave signal

译文：若用方波信号。

说明：这里 with 引导一个表示条件的介词短语作状语，意思是"若用方波信号，……"

［4］An important function of NOT gate is to produce signal inversion or an output signal that is opposite innature to the input signal.

译文：非门的重要功能是产生反相信号，或者产生与输入信号性质相反的输出信号。

说明：句中不定式 to produce signal inversion or an output signal 充当表语，其后 that 引导的定语从句修饰 an output signal。

［5］... such as a computer, control or digital-communication system, there are only a few basic operations, which must be performed.

译文：……例如，计算机系统、控制系统或数字通信系统中，需要进行的基本运算也只有几种。

说明：such as 引出 a large scale digital system 的同位语。which must be performed 是非限制性定语从句，对 a few basic operations 起补充说明作用。

Exercises

Ⅰ. **Decide whether the following statements are true (T) or False (F) according to the text.**

1. The closed-switch state can be represented by 1 and the open-switch state can be represented by 0.

2. A very common method of digital operation is achieved by using current pulses.

3. Integrated circuits containing many transistors are most commonly used as switching devices in digital electronics logic gates.

4. The three basic types of digital logic gates are the AND gate, the OR gate and the NOT gate.

5. An important function of NOT gate is to produce signal inversion or an input signal that is opposite in nature to the output signal.

6. The three basic types of digital logic gates and the flip-flop are the four circuits most commonly employed in such systems.

II. Translate the following paragraphs into Chinese.

Digital signals and circuits are the vast and important subject. Digital signals are binary in nature taking on values in one of two well-defined ranges. The set of basic operations that be performed on digital signals is quite small. The behavior of any digital system (up to and including the most sophisticated digital computer) can be represented by appropriate combinations of digital variables and the digital operations from this small set.

We concerned with digital system variables that take on only two values (binary variables). We conventionally denote these values as "0" or "1", and then use a special set of rules called Boolean algebra to summarize the various ways in which digital variables can be combined. This algebra and much of the notation are adopted directly from mathematical logic. Thus, "logic variable" or "logic operation" are commonly used in place of digital variable, or digital operation.

Unit 5 The Electronic Instruments

Text

1. Multimeters

A multimeter is a general-purpose meter capable of measuring DC and AC voltage, current, resistance, and in some cases, decibels. There are two types of meters: analog, using a standard meter movement with a needle (seen in Fig. 5-1a), and digital, with an electronic numerical display (seen in Fig. 5-1b). Both types of meters have a positive (+) jack and a common jack (-) for the test leads, a function switch to select DC voltage, AC voltage, DC current, AC current, or ohm and a range switch for accurate readings. The meters may also have other jacks to measure extended ranges of voltage (1 to 5 kV) and current (up to 10A). There are some variations to the functions used for specific meters.

Besides the function and range switches (sometimes they are in a single switch), the analog meter may have a polarity switch to facilitate reversing the test leads. The needle usually has a screw for mechanical adjust to set it to zero and also a zero adjust control to compensate for weakening batteries when measuring resistance.[1] An analog meter can read positive and negative voltage by simply reversing the test leads or moving the polarity switch. A digital meter usually has an automatic indicator for polarity on its display.

Meters must be properly connected to a circuit to ensure a correct reading. A voltmeter is always placed across (in parallel) the circuit or component to be measured. When measuring current, the circuit must be opened and the meter inserted in series with the circuit or component to be measured. When measuring the resistance of a component in a circuit, the voltage to the circuit must be removed and the meter placed in parallel with the component.

2. The Oscilloscope

The oscilloscope (seen in Fig. 5-2) is basically a graph-displaying device—it draws a graph of an electrical signal. When the signal is inputted into the oscilloscope, an electron beam is created, focused, accelerated, and properly deflected to display the voltage waveforms on the face of a cathode-ray tube (CRT).[2]

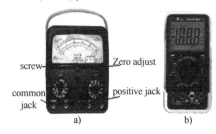

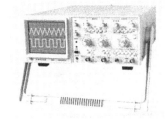

Fig. 5-1 Portable Multimeters
a) analog multimeter   b) digital multimeter

Fig. 5-2 Dual-trace Oscilloscope

In the most applications, the graph shows how signals change over time: the vertical (Y) axis represents voltage and the horizontal (X) axis represents time. The amplitude of a voltage waveform on an oscilloscope screen can be determined by counting the number of centimeters (cm), vertically, from one peak to the other peak of the waveform (seen in Fig. 5-3) and the multiplying it by the setting of the V/cm control. [3] As an example, if the amplitude was 5cm and the control was set on 1V/cm, the peak-to-peak voltage would be 5V.

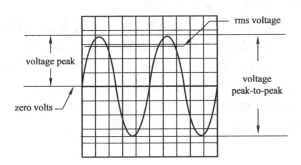

Fig. 5-3 Voltage Peak and Peak-to-peak Voltage

Time can be measured using the horizontal scale of the oscilloscope. Time measurements include measuring the period, pulse width and frequency. Frequency is the reciprocal of the period, so once you know the period, the frequency is one divided by the period.

The frequency of a waveform can be determined by counting the number of centimeters, horizontally, in one cycle of the waveform and the multiplying it by the setting time/cm control. For example, if the waveform is 4cm long and the control is set at 1ms/cm, the period would be 4ms. The frequency can now be found from the formula:

$$f = \frac{1}{T} = \frac{1}{4\text{ms}} = 250\text{Hz}$$

If the control was gone on 100μs/cm, the period would be 400μs and the frequency would be 2.5kHz.

A dual-trace oscilloscope is advantageous to show the input signal and output signal of one circuit in the same time, to determine any defects, and indicate phase relationships. The two traces may be placed over each other (superimposed) to indicate better the phase shift between two signals.

New Words and Phrases

accelerate　　[əkˈseləreit]　　*vt.* 使……加快　*vi.* 加速；促进；增加
multimeter　　[mʌlˈtimitə]　　*n.* 万用表，数字万用表，多用电表
oscilloscope　[ɔˈsiləskəup]　　*n.* 示波器
amplitude　　[ˈæmplitjuːd]　　*n.* 广阔，丰富，振幅
analog　　　[ˈænəlɔg]　　*n.* 类似物；模拟　*adj.* 有长短针的；模拟的
facilitate　　[fəsiliteit]　　*vt.* 促进；帮助；使容易
insert　　　[inˈsəːt, ˈinsəːt]　　*vt.* 插入；嵌入　*n.* 插入物
deflect　　　[diˈflekt]　　*vt.* 使偏斜；使转向；使弯曲　*vi.* 偏斜；转向
multiply　　[ˈmʌltiplai]　　*v.* 乘；(使)相乘　*adv.* 多样地　*adj.* 多层的
reciprocal　　[riˈsiprəkəl]　　*adj.* 相互的；倒数的　*n.* 倒数
beam　　　[ˈbiːm]　　*n.* 梁；(光线的)束；电波　*v.* 播送

compensate ['kɔmpenseit] v. 偿还，补偿，付报酬
decibel ['desibel] n. 分贝
defect ['diːfekt, diˈfekt] n. 缺点，缺陷 vi. 叛变；变节
focus ['fəukəs] n. 焦点；焦距；中心 v. （使）聚焦
horizontally [hɔːriˈzɔntli] adv. 水平地
jack [dʒæk] n. 插孔，插座，起重器 vt. （用起重器）抬起
meter ['miːtə] n. 仪表；米 vt. 用仪表测量 vi. 用表计量
needle ['niːdl] n. 针；指针；针状物；刺激 vi. 缝纫；做针线
period ['piəriəd] n. 周期，期间；课时 adj. 某一时代的
superimposed ['sjuːpəimˈpəuzd] adj. 上叠的；重叠的；叠加的
vertical ['vəːtikəl] adj. 垂直的 n. 垂直线

Technical Terms

cathode-ray tube (CRT)　阴极射线显像管
dual-trace oscilloscope　双踪示波器
in parallel with　与……并联
in series with　与……串联
peak-to-peak voltage　电压峰-峰值
phase shift　相位漂移/差别，移相
polarity switch　极性开关
rms voltage　电压有效值

Notes

[1] The needle usually has ascrew for mechanical adjust to set it to zero and also a zero adjust control to compensate for weakening batteries when measuring resistance.

译文：指针常常有一个旋钮用于机械调零。当测量电阻时，一个零点调节控制（钮）用来对万用表内电池电压的不足作出补偿调节（即保证电阻为 0 时指针指向零值）。

说明：句中 to compensate for weakening batteries 作后置定语。

[2] When the signal is inputted into the oscilloscope, an electron beam is created, focused, accelerated, and properly deflected to display the voltage waveforms on the face of a cathode-ray tube (CRT).

译文：当信号输入到示波器中时，一个电子束被产生、聚焦、加速并适当偏离，在阴极射线管的显示屏上显示电压的波形。

说明：句中 to display the voltage waveforms... 作结果状语。

[3] The amplitude of a voltage waveform on an oscilloscope screen can be determined by counting the number of centimeters (cm), vertically, from one peak to the other peak of the waveform (seen in Fig. 4-3) and the multiplying it by the setting of the V/cm control.

译文：示波器屏幕上电压波形的幅度，可以通过计算电压波峰与波谷之间纵向距离来确定（见图 4-3），将这个值乘以 V/cm 控制钮的设定值，就得到电压的幅度值。

说明：句中 from one peak... 作后置定语，修饰 centimeters；by counting... and the multiplying it by... 是方式状语，修饰 be determined。

Exercises

Ⅰ. **Answer the following questions according to the text.**
1. What can be measured with a multimeter?
2. What is the function of an oscilloscope?
3. Do you know the differences between a digital multimeter and an analog meter? Can you describe them?
4. What is the advantage of a dual-trace oscilloscope?

Ⅱ. **Translate the following phrases into Chinese.**
1. general-purpose meter
2. reverse the test leads
3. mechanical adjust
4. voltage amplitude
5. dual-trace oscilloscope
6. signal generator
7. analog multimeter
8. phrase relationship
9. display the voltage waveform
10. positive voltage

Ⅲ. **Translate the following sentences into English.**
1. 模拟万用表有一个极性开关，可以很方便地交换测试笔的极性。
2. 对特殊的万用表而言，还有一些其他功能的变化。
3. 示波器显示一个电子信号的图像。
4. 双踪示波器具有同时显示输入信号和输出信号的优点。
5. 两路信号的波形重叠在一起能显示输入与输出信号相位的差别。

Translating Skills

增 词 译 法

词的增译就是在译文中增加一些原文中无其形而有其义的词。英语中有时为了避免重复，常省略一些词而不影响全句意义的完整表达，但在汉语译文中如果省略了这些词，就会使译文意义不明确、不通顺。所以，在某些场合的翻译中增词法是非常必要的。

1. 用汉语动词补充英语名词、动名词或介词的意义，使译文通顺

The world needn't be afraid of a possible shortage of coal, oil, natural gas or other sources of fuel for the future. 世界无须担心将来可能出现煤、石油、天然气，或其他燃料来源短缺的问题。

The molecules get closer and closer with the pressure. 随着压力增加，分子越来越接近。

2. 在表达动作意义的英语名词后增添汉语名词

This lack of resistance in very cold metals may become useful in electronic computer. 这种在甚低温中金属没有电阻的现象，可能对电子计算机很有作用。

The lower the frequency is, the greater the refraction of a wave will be. 频率越低，波的折射作用就越强。

3. 增加表示名词复数的词

The moving parts of a machine are often oiled that friction may be greatly reduced. 机器的各个可动部件被润滑油润滑，以便大大减少摩擦。

The first electronic computers used vacuum tubes and other components and this made equipment very large and bulky. 第一代电子计算机使用电子管和其他元器件，这使设备又大又笨。

31

4. 增加某些被动语态或动名词中没有具体指出的动作执行者或暗含的逻辑主语

The material is said to behave elastically. 人们常说，这种材料具有弹性。

To explore the moon's surface, rockets were launched again and again. 为了勘探月球的表面，人们一次又一次的发射火箭。

5. 在形容词前加名词

According to Newton's Third Law of Motion, action and reaction are equal and opposite. 根据牛顿第三运动定律，作用力和反作用力是大小相等方向相反的。

The washing machine of this type is indeed cheap and fine. 这种类型的洗衣机真是物美价廉。

6. 增加表示数量意义的概括性的词，起修饰润色作用

The frequency, wavelength and speed of sound are closely related. 声音的频率、波长与速度三者密切相关。

A designer must have a good foundation in statics, kinematics dynamics and strength of materials. 一个设计人员必须在静力学、运动学、动力学和材料力学这四个方面有很好的基础。

7. 增加使译文语气连贯的词

Manganese is a hard, brittle, gray-white metal. 锰是一种灰白的、又硬又脆的金属。

In general, all the metals are good conductors, with silver the best and copper the second. 一般来说，金属都是良导体，其中，银最佳，铜次之。

Reading

How to Use a Tester

This instrument is designed to use for measuring DC voltages, measuring AC voltages, measuring resistance, conductivity test and diode test. It has $3\left(\frac{1}{2}\right)$ digits liquid crystal display. So it is called a digital multi-meter.

The following is illustrating its operation.

1. Measuring DC voltages (Seen in Fig. 5-4)

　1) Set the function switch to "DCV".

　2) Connect the test leads to the circuit to be measured.

　3) Read the display.

Note：

　1) "-" (minus sign) is displayed when the polarity of the test leads is reversed.

　2) Use the test leads with the normal polarity when measuring a voltage that includes spike pulses (such as horizontal output signal of a TV set).

2. Measuring AC Voltage (Seen in Fig. 5-5)

　1) Set the function switch to "ACV".

　2) Connect the test leads to the circuit to be measured.

　3) Read the display.

Note：It is not necessary to consider the polarity of the test leads.

Fig. 5-4　Measuring DC Voltages

Fig. 5-5　Measuring AC Voltage

3. Measuring Resistance (Seen in Fig. 5-6)

1) Set the function switch to "Ω".
2) Connect the test leads to the circuit to be measured.
3) Read the display.

Note: Be sure to turn off the power of the circuit to be measured before connecting the leads.

4. Conductivity Test (Seen in Fig. 5-7)

1) Set the function switch to " ·)) ".
2) Connect the test leads to the circuit to be tested.
3) Conductivity is good when the buzzer beeps and the mark " ·)) " is displayed.

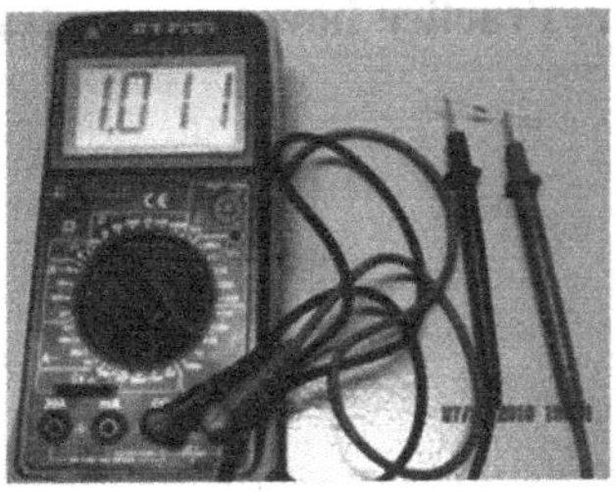

5. Diode Test (Seen in Fig. 5-8)

1) Set the function switch to " ·)) ".
2) With a normal diode, the display shows the forward resistance of the diode when the black test lead is connected to the cathode of the diode and the red test lead to the anode; it displays "1." when the test leads are reversed.

Fig. 5-6　Measuring Resistance

3) When the test leads are open, the display reads "1.".

Fig. 5-7　Conductivity Test

Fig. 5-8　Diode Test

New Words and Phrases

conductivity　[kɒndʌk'tɪvɪtɪ]　*n.* 导电性；传导性
illustrate　['iləstreit]　*vt.* 阐明，举例说明，图解
function　['fʌŋ(k)ʃ(ə)n]　*n.* 功能；函数；职责　*vi.* 运行；起作用
display　[dɪ'spleɪ]　*vt.* 显示；表现；陈列　*adj.* 展览的；陈列用的　*n.* 显示（器）

spike　［spaik］　*n.* 长钉，尖峰信号，*vt.* 用尖物刺穿
buzzer　［ˈbʌzə］　*n.* 蜂鸣器；嗡嗡作声的东西
beep　［biːp］　*vi.* 嘟嘟响　*n.* 哔哔的声音，警笛声
cathode　［ˈkæθəud］　*n.* 阴极，负极
multi-meter　万用电表，多量程仪表
DC（Direct Current）　直流
AC（Alternate Current）　交流
test leads　表笔

Exercises

I. Translate the following sentences into English.
1. 该仪表是用于测量电压、电阻和二极管的。
2. 该仪表的使用方法示例如下。
3. 表笔极性接反时，会出现"－"（负号）。
4. 把测试表笔接在被测电路上。

II. Practice how to use a tester.

Unit 6 Portable Media Player

Text

A portable media player (PMP), is a consumer electronics device that is capable of storing and playing digital media such as audio, images, video, documents, etc. The data is typically stored on a hard drive, micro drive, or flash memory (as shown in Fig. 6-1).

Fig. 6-1 Portable Media Player

Digital audio players are generally categorized by storage media:
- **Flash-based players**: These are non-mechanical solid state devices that hold digital audio files on internal flash memory. Because they are solid state and do not have moving parts, they require less battery power, and may be more resilient to hazards such as dropping or fragmentation than hard disk-based players.
- **Hard drive-based players or digital jukeboxes**: Devices that read digital audio files from a hard disk drive (HDD). These players have higher capacities as of 2010 ranging up to 500GB. At typical encoding rates, this means that tens of thousands of songs can be stored on one player. The disadvantages with these units is that a hard drive consumes more power, is larger and heavier and is inherently more fragile than solid-state storage.
- **MP3 CD/DVD players**: Portable CD players that can decode and play MP3 audio files stored on CDs. Such players are typically much less expensive than either the hard drive or flash-based players. The blank CD-R media they use is very inexpensive, typically costing less than US $0.15 per disc. These devices have the feature of being able to play standard "red book" CD-DA audio CDs. A disadvantage is that due to the low rotational disk speed of these devices, they are even more susceptible to skipping or other misread of the file if they are subjected to uneven acceleration (shaking) during playback. The mechanics of the player itself "however" can be quite sturdy, and are generally not as prone to permanent damage due to being dropped as hard drive-based players. Since a CD can typically hold only around 700MB of data, a large library will require multiple disks to contain.[1] However, some higher-end units are also capable of reading and playing back files stored on larger capacity DVD; some also have the ability to play back and display video content, such as

movies.[2] An additional consideration can be the relatively large width of these devices, since they have to be able to fit a CD.

- **Networked audio players**: Players that connect via (WiFi) network to receive and play audio. These types of units typically do not have any local storage of their own and must rely on a server, typically a personal computer also on the same network, to provide the audio files for playback.[3]
- **USB host/memory card audio players**: Players that rely on USB flash drives or other memory cards to read data.

PMPs are capable of playing digital audio, images and video. Usually, a color-liquid crystal display (LCD) or organic light-emitting diode (OLED) screen is used as a display. Various players include the ability to record video, usually with the aid of optional accessories or cables, and audio, with a built-in microphone or from a line out cable or FM tuner. Some players include readers for memory cards.

New Words and Phrases

portable	[ˈpɔːtəbl]	adj. 便携式的 n. 便携式设备
device	[diˈvais]	n. 仪器；策略；商标图案
digital	[ˈdidʒitəl]	adj. 数字的；手指的 n. 数字
battery	[ˈbætəri]	n. 电池；一组，一套；炮台，炮位
disc	[disk]	n. 磁盘；唱片；圆盘
feature	[ˈfiːtʃə]	n. 特点 vt. 以……为特色 vi. 起重要作用
standard	[ˈstændəd]	n. 标准；度量衡标准 adj. 标准的
uneven	[ˌʌnˈiːvən]	adj. 不均匀的；不平坦的；[数学] 奇数的
multiple	[ˈmʌltipl]	adj. 多个的 n. 倍数；[电工学] 并联
organic	[ɔːˈgænik]	adj. 有机的；(动、植物的) 器官的；有组织的
tuner	[ˈtjuːnə]	n. [无线电] (高频) 调谐器，调谐设备；调音师
server	[ˈsɜːvə]	n. 服务器；发球员；侍应生，服务员
hard drive	硬驱，硬盘驱动器	
flash memory	闪存	
memory card	存储卡	

Notes

[1] Since a CD can typically hold only around 700MB of data, a large library will require multiple disks to contain.

译文：由于CD通常只能容纳大约700MB的数据，一个大型文库往往需要多个磁盘来容纳。

说明：这里 since 表示"由于，因为"，引导原因状语从句；since 还可以表示"自从……以来；自……以后"，引导时间状语从句，例如，I have distrusted her ever since she cheated me. 自从她骗了我以后，我就不信任她了。

[2] However, some higher-end units are also capable of reading and playing back files stored on larger capacity DVD; some also have the ability to play back and display video content, such as movies.

译文：然而，一些较高端的播放机也能读取和播放那些存储在容量更大的 DVD 中的文件；有的播放机还能播放和显示视频内容，如电影。

说明：句中 stored on larger capacity DVD 过去分词短语作定语修饰 files；to play back and display video content 是不定式短语作定语修饰 ability。

[3] These types of units typically do not have any local storage of their own and must rely on a server, typically a personal computer also on the same network, to provide the audio files for playback.

译文：这种类型的播放器一般没有自己的存储器，需要依靠一台服务器，通常为一台连接在同一网络上的个人计算机，来提供音频文件进行播放。

说明：这是一个简单句。主语中心词为 These types of units；do not have 与 must rely on 是并列谓语动词；a personal computer 是 a server 的同位语；also on the same network 是介词短语作后置定语；to provide the audio files for playback 是不定式作状语表示目的。

Exercises

Ⅰ. Answer the following questions according to the text.
1. What is a PMP?
2. Where is the data typically stored?
3. Which are digital audio players generally categorized?
4. What is usually used as a display when PMPs play digital audio, images and video?

Ⅱ. Fill in the missing words according to the text.

PMPs are capable of playing digital audio, images and video. Usually, a color-liquid crystal display (LCD) or organic light-emitting diode (OLED) screen is used as a _____. Various players include the ability to _____ video, usually with the aid of optional accessories or cables, and audio, with a built-in _____ or from a line out cable or FM tuner. Some players include _____ for memory cards.

Ⅲ. Choose the correct translation for the following terms.
1. PMP (Portable Media Player)
2. HDD (Hard Disk Drive)
3. CD-DA (Compact Disc Digital Audio)
4. LCD (Liquid Crystal Display)
5. OLED (Organic Light-Emitting Diode)

A. 发光二极管　　　　B. 便携式媒体播放器　　　C. 液晶显示器
D. 数字音频光碟　　　E. 硬盘驱动器

Ⅳ. Translate the following sentences into Chinese.
1. Our industry has leapt forward since we followed the policy of reform and opening.
2. Digital audio players are generally categorized by storage media.
3. PMPs are capable of playing digital audio, images and video.
4. You can categorize and tag your findings for later, and keep your stuff private, shared with friends, or make it public.

Translating Skills

减词译法

英译汉时，英文中有些词语不必译出，而句意仍然清楚，这种译法称为减词译法。减词的原则是在使汉语译文完全表达英语原文内容的前提下，把译文写得简明扼要、通顺流畅，读起来就像地道的汉语一样。减词译法有以下几种情况。

1. 省略代词

By the word "alloy" we mean "mixture of metals". 用"合金"这个词来表示"金属的混合物"。

It is now clear why the laser usually oscillates on the R_1 line. 现在清楚了，为什么激光总是在 R_1 线上振荡。

2. 省略冠词

Satellites can be sent into space with the help of rockets. 借助火箭可以把卫星送上太空。

3. 省略连词

If there were no heat-treatment, metals could not be made so hard. 没有热处理，金属就不会变得如此硬。

Give him an inch and he will take a mile. 他得寸进尺。

4. 省略介词

The search for even better magnetic material is a part of the modern frontier of physics. 研制更好的磁性材料是现代物理学前沿之一。

The unit of measurement of the pressure of water is Pa. 水压的计量单位是帕。

5. 省略逻辑上或修辞上不需要的词

As we know, electrons revolve about the nucleus, or center of an atom. 正如我们所知，电子围绕着原子核旋转。

A generator can not produce energy, what it does is to convert mechanical energy into electrical energy. 发电机不能产生能量，它只能把机械能转变成电能。

Reading

Digital Audio

Digital Audio is the representation of audio information in digital (discrete level) formats. The use of digital audio allows for simple transmission, storage, and processing of audio signals.

Audio digitization is the conversion of analog audio sounds into digital form. To convert an analog audio signal to digital form, the analog signal is digitized by using an analog-to-digital (pronounced A to D) converter. The A/D converter periodically senses (samples) the level of the analog signal and creates a binary number or series of digital pulses that represent the level of the signal (as shown in Fig. 6-2). The typical sampling rate for the conversion of analog audio ranges from 8000 samples per second (for telephone quality) to 44,000 samples per second (for music quality).

Audio compression is the analysis and processing of digital sound to a form that reduces the space required for transmission or storage. Audio compression coders and decoders analyze digital audio signals to remove signal redundancies and sounds that cannot be heard by humans.[1]

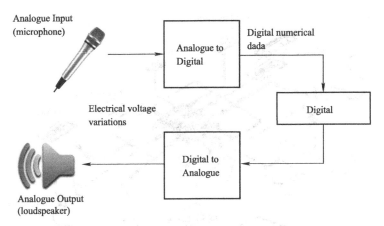

Fig. 6-2　A/D converter and D/A converter

  Digital audio data is random in nature unlike digital video which has repetitive information that occurs on adjacent image frames. This means that audio signals do not have a high amount of redundancy, making traditional data compression and prediction processes ineffective at compressing digital audio. It is possible to highly compress digital audio by removing sounds that can be heard or perceived by listeners through the process of perceptual coding. [2]

  Audio coding is the type of coder (type of analysis and compression) which can dramatically vary and different types of coders may perform better for different types of audio sounds (e. g. speech audio as compared to music).

  The MPEG system allows for the use different types of audio coders. The type of coder that is selected can vary based on the application (such as playing music or speech) and the type of device the audio is being played through (such as a television or a battery operated portable media player). [3] The MPEG speech coders range from low complexity (layer 1) to high complexity (layer 3). A new version of audio coder has been created (advanced audio codec-AAC) that offers better audio quality at lower bit rates. The AAC coder also has several variations that are used in different types of applications (e. g. broadcast radio -vs. - real time telephony).

  Multichannel audio is the use of multiple sound channels to produce an enhanced listening experience. Examples of multichannel audio include stereo, surround sound and low frequency enhancement (LFE). The MPEG system was designed to allow the combination of multiple audio channels (such as stereo).

  Many types of audioapplications can benefit from the high quality, power efficiency, space savings and ease of use in TI's all-digital audio solution. These include consumer entertainment such as HTIB, DVD receivers and mini/micro systems, PC gaming and entertainment applications, business applications, such as speakerphones and multimedia conferencing systems, and digital headsets for all application areas. In addition, TI's all-digital audio solution will enhance automotive entertainment, reducing space and heat dissipation for manufacturers while allowing consumers to enjoy the experience of high-quality multichannel digital audio in their vehicles (as shown in Fig. 6-3).

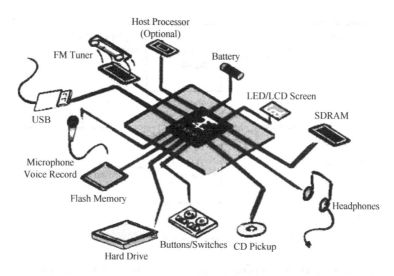

Fig. 6-3　Digital Audio System

New Words and Phrases

  analogue　　　['ænəlɔg]　　　n. 模拟；类似物，相似物　　adj. 类似的
  binary　　　['bainəri]　　adj. 二进制的；由两部分构成的　　n. 二进制数
  pulse　　　[pʌls]　　n. 脉冲；脉搏　vi. 跳动　vt. 使跳动
  coder　　　['kəudə]　　n. 编码器；编码员
  decoder　　　[ˌdiː'kəudə]　　n. 解码器，译码器；译码员
  redundancy　　　[ri'dʌndənsi]　　n. 冗余；裁员
  random　　　['rændəm]　　adj. 随机的；任意的
  prediction　　　[pri'dikʃən]　　n. 预言；预测；预报
  perceptual　　　[pə'septʃuəl]　　adj. 感知的；[心理学] 知觉的
  layer　　　['leiə]　　n. 层；[植物学] 压条　vi. 分层堆积
  bit　　　[bit]　　n. 比特；一点；少量 [常与 a 连用，起副词的作用]
  speakerphone　　　['spi'kəfəun]　　n. 扬声电话
  headset　　　['hedset]　　n. 耳机
  automotive　　　[ˌɔːtə'məutiv]　　adj. 汽车的，机动车的
  vehicle　　　['viːikl]　　n. 机动车辆；媒介物

Notes

[1] Audio compression coders and decoders analyze digital audio signals to remove signal redundancies and sounds that cannot be heard by humans.

译文：音频压缩编码器和解码器（编解码器）分析数字音频信号，去除信号冗余和人类听不到的声音。

说明：句中 to remove signal redundancies and sounds 是不定式短语作目的状语，之后的关系代词 that 引导的定语从句修饰 sounds。

[2] It is possible to highly compress digital audio by removing sounds that can be heard or

perceived by listeners through the process of perceptual coding.

译文：通过感知编码的过程，去除听者可以听到或感知的声音，高度压缩数字音频是可行的。

说明：句中 It 是形式主语，真正的主语是不定式短语 to highly compress digital audio。

[3] The type of coder that is selected can vary based on the application (such as playing music or speech) and the type of device the audio is being played through (such as a television or a battery operated portable media player).

译文：基于应用（如播放音乐或演讲）和播放音频设备的类型（如电视或使用电池的便携式媒体播放器），选择的编码器类型可以不同。

说明：这是一个复合句。The type of coder 是主句主语；that is selected 是定语从句修饰 type；can vary 是主句谓语；based on... 是过去分词短语作原因状语；句中 the audio is being played through 是省略了关系代词的定语从句修饰 device；关系代词在定语从句中作介词 through 的宾语省略。

Exercises

Ⅰ. **Answer the following questions according to the text.**

1. What is audio digitization?
2. What is audio compression?
3. What is multichannel audio?

Ⅱ. **Fill in the missing words according to the text.**

Digital audio data is _____ in nature unlike digital video _____ has repetitive information that _____ on adjacent imageframes. This means that audio signals do not have a high _____ redundancy, making traditional data compression and prediction processes _____ at compressing digital audio. It is possible to highly compress digital audio by removing sounds that can be heard or _____ by listeners through the process of _____ coding.

Chapter II Communication Technology

Unit 7 Optical Fiber Communications

Text

Communication may be broadly defined as the transfer of information from one point to another. When the information is to be conveyed over any distance, a communication system is usually required. [1] Within a communication system the information transfer is frequently achieved by superimposing or modulating the information on to an electromagnetic wave which acts as a carrier for the information signal. [2] This modulated carrier is then transmitted to the required destination where it is received and the original information signal is obtained by demodulation. Sophisticated techniques have been developed for this process by using electromagnetic carrier waves operating at radio frequencies as well as microwave and millimeter wave frequencies. [3] However, "communication" may also be achieved by using an electromagnetic carrier which is selected from the optical range of frequencies.

Typical optical fiber communications system is shown in Fig. 7-1. In this case the information source provides an electrical signal to a transmitter comprising an electrical stage which drives an optical source to give modulation of the light wave carrier. The optical source which provides the electrical-optical conversion may be either a semiconductor laser or light emitting diode (LED). The transmission medium consists of an optical fiber cable and the receiver consists of an optical detector which drives a further electrical stage and hence provides demodulation of the optical carrier. Photodiodes and, in some instances, phototransistors and photoconductors are utilized for the detection of the optical signal and the optical-electrical conversion. Thus there is a requirement for electrical interfacing at either end of the optical link and at present the signal processing is usually performed electrically.

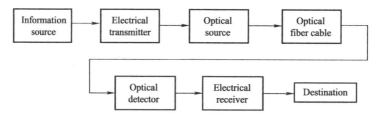

Fig. 7-1 Optical Fiber Communications System

The optical carrier may be modulated by using either an analog or digital information signal. Analog modulation involves the variation of the light emitted from the optical source in a continuous manner. With digital modulation, however, discrete changes in the light intensity are obtained

(i. e. on-off pulses). Although often simpler to implement, analog modulation with an optical fiber communication system is less efficient, requiring a far higher signal to noise ratio at the receiver than digital modulation. Also, the linearity needed for analog modulation is not always provided by semiconductor optical source, especially at high modulation frequencies. For these reasons, analog optical fiber communications link are generally limited to shorter distances and lower bandwidths than digital links.

Initially, the input digital signal from the information source is suitably encoded for optical transmission. The laser drive circuit directly modulates the intensity of the semiconductor laser with the encoded digital signal. Hence a digital optical signal is launched into the optical fiber cable. The avalanche photodiode detector (APD) is followed by a front-end amplifier and equalizer or filter to provide gain as well as linear signal processing and noise bandwidth reduction. [4] Finally, the signal obtained is decoded to give the original digital information.

New Words and Phrases

optical ['ɔptikəl] *adj.* 光学的；眼睛的，视觉的
fiber ['faibə] *n.* 纤维；光纤（等于 fibre）
convey [kən'vei] *v.* 传达，运输，转移；输送
superimpose ['sjuːpərim'pəuz] *vt.* 重叠（安装，添加）
modulate ['mɔdjuleit] *v.* 调整，转调
demodulation [diːmɔdju'leiʃən] *n.* 检波；反调制；解调制
sophisticated [sə'fistikeitid] *adj.* 富有经验的；精致的；复杂的
photoconductor *n.* 光电导体；光电导元件；光敏电阻
phototransistor [ˌfəutəutræn'zistə] *n.* 光敏晶体管
be utilized for 用于
detection [di'tekʃən] *n.* 察觉；侦查，探测；发觉，发现
manner ['mænə] *n.* 方式；习惯；种类；规矩；风俗
launch into 进入，投入
equalizer ['iːkwəlaizə] *n.* 平衡装置；均衡器，平衡器；使相等的东西

Technical Terms

optical fiber 光纤，光导纤维
modulated carrier 调制载波，被调载波，已调载波
communication system 通信系统
optical detector 光检测器；光学探测器；光辐射探测器
optical signal 光信号
analog modulation 模拟调制
digital modulation 数字调制
light intensity 光强度
APD (*abbr.* avalanche photodiode detector) 雪崩光电探测器
front-end amplifier 前置放大器

Notes

[1] When the information is to be conveyed over any distance, a communication system is usually required.

译文：当要跨越一段距离传送信息时，就需要一个通信系统。

说明：to be conveyed 是动词不定式的被动语态，作 the information 的表语，表示主语的所处的状态；由 when 引导的句子为状语从句，表示动词 require（被动语态 be required）发生的时间。

[2] Within a communication system the information transfer is frequently achieved by superimposing or modulating the information on to an electromagnetic wave which acts as a carrier for the information signal.

译文：在通信系统中，信息传送时常是通过把信息叠加在电磁波上或对电磁波进行调制来实现的，这样的电磁波起着传送信号的作用。

说明：主句 the information transfer is frequently achieved 使用了被动语态；by superimposing... 是介词短语，在句子中起状语的作用，表示动作的方式；superimposing or modulating 是动名词，充当介词 by 的宾语；which acts as a carrier for the information signal 是定语从句，修饰 electromagnetic wave。

[3] Sophisticated techniques have been developed for this process by using electromagnetic carrier waves operating at radio frequencies as well as microwave and millimeter wave frequencies.

译文：用无线电频率及微波和毫米波的频率作为载波频率的技术已很成熟。

说明：by using electromagnetic carrier waves operating at radio frequencies as well as microwave and millimeter wave frequencies 中，介词 by + 动名词短语充当状语；as well as 意思是"以及"，相当于 and，常用于连接并列成分。

[4] The avalanche photodiode detector (APD) is followed by a front-end amplifier and equalizer or filter to provide gain as well as linear signal processing and noise bandwidth reduction.

译文：雪崩光敏二极管（APD）检波器后有前置放大器和均衡器或滤波器，用来提供增益，对信号进行线性处理和减少噪声带宽。

说明：to provide... 为不定式短语，充当目的状语。

Exercises

I. Answer the following questions according to the text.

1. How is the information transfer achieved?
2. Which parts does an optical fiber communication system consist of?
3. Why are generally analog optical fiber communications link limited to shorter distances and lower bandwidths than digital links?
4. Why is the input digital signal from the information source encoded suitably for optical transmission?
5. How is the information transfer achieved within a Digital Optical Fiber Communications System?

II. Fill in the missing words according to the text.

1. Within a communication system the information transfer is _____ by superimposing or modulating the information on to an electromagnetic wave which acts as a carrier.
2. Photodiodes and phototransistors and photoconductors are utilized for the _____ of the

optical signal and the optical-electrical _____.

3. The laser drive circuit directly modulates the intensity of the semiconductor laser with the encoded digital signal. Hence a _____ is launched into the optical fiber cable.

4. Finally, the signal obtained is _____ to give the original digital information within an optical fiber communication system.

5. There is a requirement for electrical interfacing at either end of the optical link and at present the _____ is usually performed electrically.

Ⅲ. **Choose the best answer for the following sentences.**

1. This modulated carrier is then transmitted to the required destination where it is _____ and the original information signal is obtained by modulation.
   A. received B. transmitted C. emitted

2. Sophisticated techniques have been developed for this process by using magnetic carrierwaves operating at radio frequencies _____ microwave and millimeter wave frequencies.
   A. as well as B. and C. or

3. The information source provides an electrical signal to a transmitter _____ an electrical stage which drives an optical source to give modulation of the light wave carrier.
   A. comprising B. comprised C. to consist

4. The optical carrier can be modulated by using _____.
   A. digital information signal B. analog information signal C. A or B

5. Analog modulation with an optical fiber communication system is _____ efficient, requiring a far _____ signal to noise ratio at the receiver than digital modulation.
   A. less, higher B. more, higher C. more, lower

6. The input digital signal from the information source is suitably _____ for optical transmission.
   A. encoded B. decoded C. launched

Ⅳ. **Translate the following sentences into English.**

1. 把信息从一个地方传递到另一个地方就称为通信。
2. 当要跨越一段距离传送信息时，就需要一个通信系统。
3. 在通信系统中，信息传送时常是通过把信息叠加在电磁波上或对电磁波进行调制来实现的，这样的电磁波起着传送信号的作用。
4. 最后，解码得到原始信号。

Translating Skills

科技英语词汇的结构特征（Ⅰ）

科技英语词汇的构成主要有派生法、转化法、合成法、缩略法和类比造词法。

1. 派生法（derivation）

派生法也叫词缀法（affixation）。在一个词根（或词干）上添加构词词缀构成词的方法，称为派生法。常见的派生法分为加后缀、加前缀和加前、后缀三类。

常用前缀：thermo-表示"热"   如：thermo-dynamics（热力学）
　　　　　micro-表示"微"   如：microchip（微芯片）
　　　　　super-表示"超、过"   如：superconductor（超导体）

	hydro- 表示"水、氢"	如：hydropathy（水疗法）
常用后缀：	-scope 表示"仪器"	如：spectroscope（分光仪）
	-graph 表示"记录工具"	如：chromatograph（色谱仪）
	-meter 表示"计、仪"	如：spectrometer（分光计）

2. 转化法（convertion）

一个单词由某个词性转化成另一个词性而词形不变，这种构词法称为转化法，主要有名词转化成动词、形容词转化成动词、动词转化成名词、形容词转化成名词等几种类型。

名词转化成动词，如：alloy（合金）→ to alloy（制成合金）

形容词转化成动词，如：round（圆的）→ to round（使成圆形）

动词转化成名词，如：to divide（分割）→ divide（分水界）

3. 合成法（compounding）

合成法是科技术语形成和扩展的最重要途径。一般来说，合成法主要构成复合名词，其次是复合形容词。在科技英语中，由于科技术语的特性，合成法主要构成的是复合名词，其主要结构特点是"修饰成分+被修饰成分"，拼写方式有合写式（无连字符）与分写式（带连字符）两种，如：

合写式		分写式	
fallout	放射性尘埃	hot-press	热压
sunspot	太阳黑子	program-control	程序控制
windscreen	挡风屏	pulse-scaler	脉冲定标器
landslide	山崩	cross-breed	杂交合词法

Reading

Wireless Communications

Wireless is a term used to describe telecommunications in which electromagnetic waves (rather than some form of wire) carry the signal over part or the entire communication path. The first wireless transmitters went on the air in the early 20th century using radio telegraphy. Later, as modulation made it possible to transmit voices and music via wireless, the medium came to be called "radio". With the advent of television, fax, data communication, and the effective use of a larger portion of the spectrum, the term "wireless" has been resurrected.

Wireless can be divided into several kinds:

- Fixed wireless: the operation of wireless devices or systems in homes and offices, and in particular, the equipment connected to the Internet via specialized modems.
- Mobile wireless: the use of wireless devices or systems aboard, moving vehicle. Examples include the automotive cell phone and PCS (personal communications services).
- Portable wireless: the operation of autonomous, battery-powered wireless devices or systems outside the office, home or vehicle. Examples include hand-held cell phones and PCS units.
- IR wirelcss: the use of devices that convey data via IR radiation, employed in certain limited-range communications and control systems.
- Common examples of wireless equipment in use today include the following:
- Cellular phones and pagers: provide connectivity for portable and mobile applications both

personal and business.
- Global Positioning System (GPS): allows drivers of cars and trucks, captains of boats and ships, and pilots of aircraft to ascertain their location anywhere on earth.
- Cordless computer peripherals: the cordless mouse is a common example; keyboards and printers can also be linked to a computer via wireless.
- Cordless telephone sets: these are limited-range devices, not to be confused with cell phones.
- Home-entertainment-system control boxes: the VCR control and the TV channel control are the most common examples; some hi-fi sound systems and FM broadcast receivers also use this technology.
- Remote garage-door openers: one of the oldest wireless devices in common use by consumers, usually operating at radio frequencies.
- Two-way radios: this includes Amateur and Citizens Radio Service, as well as business, marine and military communications.
- Baby monitors: these devices are simplified radio transmitter or receiver units with limited range.
- Satellite television: allows viewers in almost any location to select from hundreds of channels.
- Wireless LANs: provide flexibility and reliability for business computer users.

Wireless technology is rapidly evolving, and is playing an increasing role in the lives of people throughout the world. In addition, ever-larger numbers of people are relying on the technology directly or indirectly. More specialized and wonderful examples of wireless communications and control include:
- Global System for Mobile Communication (OSM): a digital mobile telephone system used in Europe and other parts of the world; the de facto wireless telephone standard in Europe.
- General Packet Radio Service (GPRS): a packet-based wireless communication service that provides continuous connection to the Internet for mobile phone and computer users.
- Enhanced Data GSM Environment (EDGE): a faster version of the Global System for Mobile (GSM) wireless service.
- Universal Mobile Telecommunications System (UMTS): a broadband, packet-based system offering a consistent set of services to mobile computer and phone users no matter where they are located in the world.
- Wireless Application Protocol (WAP): a set of communication protocols to standardize the way that wireless devices, such as cellular telephones and radio transceivers, can be used for Internet access.
- i-Mode: the world's first "smart phone" for Web browsing, first introduced in Japan, it provides color and video over telephone sets.

New Words and Phrases

telecommunications ['telɪkəˌmjuːnɪ'keɪʃənz] n. 无线电通信，电信；远程通信
electromagnetic [ɪˌlektrə(ʊ)mæg'netɪk] adj. 电磁的；电磁学的
advent ['ædvənt] n. 出现，到来

resurrect   [rezə'rekt]   *vt.* 使复活；复兴；挖出 *vi.* 复活
autonomous   [ɔː'tɔnəməs]   *adj.* 独立的，自治的
IR = infra-red   [ˌinfrə'red]   *adj.* 红外线的红外线
ascertain = find out   [ˌæsə'tein]   *vt.* 确定；查明；探知
marine   [mə'riːn]   *adj. & n.* 航海的，船舶的；舰队，海运业
de facto   [diː'fæktəu]   *adj.* 事实上（的），实际上（的）
consistent   [kən'sistənt]   *adj.* 一致的，相容的，调和的
WAP（Wirless Application Protocol） 无线电应用协议信号
battery-powered   电池驱动的
VCR control   录像机遥控器
hi-fi sound systems   高保真立体声音响系统
FM broadcast receiver   调频广播接收器
Amateur and Citizens Radio Service   业余爱好者和市民无线电服务设备
GSM（Global System for Mobile Communication） 全球数字移动电话系统
GPRS（General Packet Radio Service） 通用包无线电服务程序
EDGE（Enhanced Data GSM Environment） 增强型数据全球数字移动电话系统环境
UMTS（Universal Mobile Telecommunications System） 通用移动电信系统
i-Mode   日本 1999 年推出的世界第一个智能电话

Exercises

I. Answer the following questions according to the text.

1. What's the meaning of wireless communication?
2. What's the sort of the wire?
3. What is the purpose of the GPRS?

II. Translate the following phrases and expressions.

1. Wireless LANs
2. Cordless telephone sets
3. GSM
4. WAP

Unit 8　Satellite Communications

Text

   Satellite communication has become a part of everyday life in the late 1980s. An international telephone call is made as easily as a local call to a friend who lives down the block. We also see international events, such as an election in England and a tennis match in France, with the same regularity as local political and sporting events. In this case, a television news program brings the sights and sounds of the world into our homes each night.

   This capability to exchange information on a global basis, be it a telephone call or a news story, is made possible through a powerful communications tool—the satellite. [1] For those of us who grew up at a time when the space age was not a part of everyday life, satellite-based communication is the culmination of a dream that stretches back to an era when the term satellite was only an idea conceived by a few inspired individuals. [2] These pioneers included authors such as Arthur C. Clarke, who fostered the idea of a worldwide satellite system in 1945. This idea has subsequently blossomed into a sophisticated satellite network that spans the globe.

   The latter type of satellite system would have entailed the development of a very complex and cumbersome earth and space-based network. Fortunately though, this problem was eliminated in 1963 and 1964 through the launching of the Syncom satellite. Rather than circling the earth at a rapid rate of speed, the spacecraft appeared to be stationary or fixed in the sky. Today's communications satellites, for the most part, have followed suit and are now placed in what are called geostationary orbital positions or "slots". [3]

   Simply stated, a satellite in a geostationary orbital position appears to be fixed over one portion of the earth. At an altitude of 22,300 miles above the equator, a satellite travels at the same speed at which the rotates, and its motion is synchronized with the earth's rotation. Even though the satellite is moving at an enormous rate of speed, it is stationary in the sky in relation to an observer on the earth.

   The primary value of a satellite in a geostationary orbit is its ability to communicate with ground stations in its coverage area 24 hours a day. This orbital slot also simplifies the establishment of the communications link between a station and the satellite. Once the station's antenna is properly aligned, only minor adjustments may have to be made in the antenna's position over a period of time. The antenna is repositioned to a significant degree only when the station establishes contact with a satellite in a different slot. Prior to this era, a ground station's antenna had to physically track a satellite as it moved across the sky.

   Based on these principles, three satellites placed in equidistant positions around the earth can create a world-wide communications system in that almost every point on the earth can be reached by satellite (as shown in Fig. 8-1). This concept was the basis of Arthur Clarke's original vision of a globe-spanning communications network.

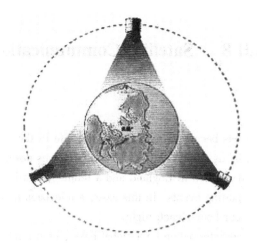

Fig. 8-1 Satellite Communication

New Words and Phrases

election　　[iˈlekʃən]　　n. （投票）选举，推举；当选；选择权
capability　　[ˌkeipəˈbiləti]　　n. 能力；容量；可能，可能性
basis　　[ˈbeisis]　　n. 基础；（物体的）底部；基本原则
conceive　　[kənˈsiːv]　　vt. 想到；想象；以为
culmination　　[kʌlmɪˈneɪʃ(ə)n]　　n. 顶点；高潮
pioneer　　[ˌpaiəˈniə]　　n. 先驱者；[美] 拓荒者　vt. 开拓
foster　　[ˈfɔstə]　　vt. 怀抱，抱有（希望等），心怀；培养；养育
subsequently　　[ˈsʌbsikwəntli]　　adv. 随后，后来
blossom　　[ˈblɔsəm]　　vi. 发展；开花　n. （尤指果树的）花
sophisticated　　[səˈfistikeitid]　　adj. 复杂的；老于世故的；富有经验的
stationary　　[ˈsteiʃənəri]　　adj. 静止的，不动的；固定的；定居的
synchronize　　[ˈsiŋkrənaiz]　　vt. 使同步　vi. 同步；同时发生
rotation　　[rəuˈteiʃən]　　n. [天文学] 自转；旋转；轮流
antenna　　[ænˈtenə]　　n. 天线；[动物学] 触角，触须
track　　[træk]　　vt. （用望远镜、雷达等）跟踪　n. 足迹；轨道
satellite communication　　卫星通信
Syncom satellite　　同步通信卫星
prior to　　在前，居先，先于
globe-spanning communications network　　全球通信网络

Notes

[1] This capability to exchange information on a global basis, be it a telephone call or a news story, is made possible through a powerful communications tool— the satellite.

译文：这种进行全球性信息交流的能力，无论是电话还是新闻转播都是通过一个强有力的通信工具——卫星，才得以实现的。

说明：本句是复合句。be it a telephone call or a news story 是一个省略了 no matter what 的

虚拟条件句，This capability 是主句主语中心词，其后面 to exchange information... 是不定式短语作后置定语，修饰 capability，is made possible 是主句谓语。

[2] For those of us who grew up at a time when the space age was not a part of everyday life, satellite-based communication is the culmination of a dream that stretches back to an era when the term satellite was only an idea conceived by a few inspired individuals.

译文：对于我们中那些并非生长在太空时代的人们来讲，卫星通信是人们长期以来一种梦想的顶点，这个梦想可以一直追溯到卫星这个词只是几个天才头脑中灵感的想象那个年代。

说明：这也是一个复合句。For those of us 是范围状语；who grew up at a time when... 是一个复合定语从句，关系副词 when 引导的定语从句修饰 time，而 who 引导的定语从句修饰 those；communication 是主句主语中心词，其后是系表结构作谓语，that 引导的定语从句修饰 dream，第 2 个 when 引导定语从句修饰 era。

[3] Today's communications satellites, for the most part, have followed suit and are now placed in what are called geostationary orbital positions or "slots".

译文：目前大部分通信卫星都被定位在相对地球静止的轨道，或称之为"槽"的位置上。

说明：本句是复合句。satellites 是主句主语，have followed 与 are placed 构成并列谓语，句中 what 引导宾语从句作介词 in 的宾语。

Exercises

I. Answer the following questions according to the text.

1. Whenhas satellite communication become a part of everyday life?
2. Who fostered the idea of a worldwide satellite system in 1945?
3. What is the primary value of a satellite in a geostationary orbit?
4. How manysatellites placed in equidistant positions around the earth can create a world-wide communications system?

II. Choose the correct translation for the following expressions.

1. satellite communication　　2. Syncom satellite
3. rotation　　4. globe-spanning communications network

　A. 全球通信网络　　B. 自转
　C. 同步通信卫星　　D. 卫星通信

III. Fill in the missing words according to the text.

The primary value of a satellite in a geostationary orbit is its ability to———＿＿＿ ground stations in its coverage area 24 hours a day. This orbital slot also simplifies the establishment of the communications link between a ＿＿＿ and the satellite. Once the station's ＿＿＿ is properly aligned, only minor adjustments may have to be made in the antenna's position over a period of time. The antenna is repositioned to a significant degree only when the station establishes contact with a satellite in a different slot. Prior to this era, a ground station's antenna had to physically ＿＿＿ a satellite as it moved across the sky.

IV. Translate the following sentences into Chinese.

1. This idea has subsequently blossomed into a sophisticated satellite network that spans the globe.

2. Fortunately, learning and learning to learn well, will always be one of the most valuable skills in your personal and professional life.

3. That is a marvelous culmination that can be achieved in no other way.

4. To determine the scope of and effort required for a project, you should verify organizational readiness prior to making any recommendations.

Translating Skills

科技英语词汇的结构特征（Ⅱ）

4. 缩略法（shortening）

通过将原词缩短或将原词包含的成分加以压缩、省略的办法构成新词，称为缩略法。缩略词就结构来说可以分为截短词、混成词和字母缩略词。

1）截短词（clippings）主要有3种情况：截头、去尾、既截头又去尾。

截头：helicopter→copter 直升（飞）机

去尾：microphone→mike 扩音器、传声器

既截头又去尾：influenza→flu 流行性感冒

2）混成词（blends），即两个词各取一部分合在一起构成一个新词，有4种构成方式。

① 取第一个词的首部接第二个词的尾部，如：

pictogram（图像电信） - picture + telegram

② 保持第一个词的原形，删去第二个词的首部，如：

lunarnaut（登月宇宙航行员） - lunar + astronaut

③ 保持第二个词的原形，删去第一个词的尾部，如：

helilift（用直升飞机运送） - helicopter + lift

④ 删去第一个词和第二个词的尾部，如：

comsat（通信卫星） - communications + satellite

3）字母缩略词（acronyms）。科技英语词汇中，字母缩略词有两种构成方式。

① 首字母缩略词，即将词语中主要词语的首字母联合成一个新词，如：

AIDS = Acquired ImmunodeficiencySyndrome（艾滋病）

CAI = Computer Assisted Instruction（计算机辅助教学）

AC = Alternating Current（交流电）

② 取单词的第一个字母和单词中的另外一个字母或者部分字母形成新的科技英语词，如：

DNA = deoxyribonucleicacid（脱氧核糖核酸） TB = tuberculosis（肺结核）

5. 类比造词法（analogical creation）

类比造词法是指通过仿照原有的同类词创造出与之对应的词或近似词的构词方式。这种构词手段反映出人们对事物某些特点的认识、比较、联想，因此主要属语义构词，其主要语体特点是生动形象、含蓄幽默、方便易懂。例如，broadcast（广播）一词是在1921年才出现的，此后"-cast"出现在许多词中，如 radiocast（无线电广播），newscast（新闻广播）等。英语新词中也有一些是通过类比法构成的。如由 hijack（劫机）类比造出了 carjack（劫车），由 citizen（市民）类比造出了 netizen（网民）等。

除以上几种主要的构词法外，科技英语还利用逆序法（back-formation）构词，如从 typewriter 逆生出 type - write，从 diagnosis 逆生出 diagnose。另外，许多科技名词和术语是借

用专有名词（包括人名、地名、商标和机构等）来造词的。如 Mace（迈斯神经镇静剂）原来是商标，现在这个商标不仅被医学采用，而且还派生出了动词 to mace（喷以迈斯神经刺激剂）；uranium（铀）来自星体 Uranus（天王星）的名字。

Reading

Internet Telephony & VoIP (Voice over Internet Protocol)

  Internet telephony consists of a combination of hardware and software that enables you to use the Internet as the transmission medium for telephone calls. For users who have free, or fixed-price Internet access, Internet telephony software essentially provides free telephone calls anywhere in the world. In its simplest form, PC-to-PC Internet telephony can be as easy as hooking up a microphone to your computer and sending your voice through a cable modem to a person who has Internet telephony software that is compatible with yours. [1] This basic form of Internet telephony is not without its problems. However, connecting in this way is slower than using a traditional telephone, and the quality of the voice transmissions is also not near the quality you would get when placing a regular phone call.

  Many Internet telephony applications are available. Some, such as CoolTalk and NetMeeting, come bundled with popular Web browsers. Others are stand-alone products. Internet telephony products are sometimes called IP telephony, Voice over the Internet (VoI) or Voice over IP (VoIP) products.

  VoIP is another Internet-based communications method which is growing in popularity. VoIP hardware and software work together to use the Internet to transmit telephone calls by sending voice data in packets using IP rather than by traditional circuit transmissions, called PSTN (Public Switched Telephone Network). The voice traffic is converted into data packets then routed over the Internet, or any IP network, just as normal data packets would be transmitted. When the data packets reach their destination, they are converted back to voice data again for the recipient. Your telephone is connected to a VoIP phone adapter (considered the hardware aspect). This adapter is connected to your broadband Internet connection. The call is routed through the Internet to a regular phone jack, which is connected to the receiver's phone. Special hardware (the phone adapter) is required only for the sender.

  Much like finding an Internet service provider (ISP) for your Internet connection, you will need to use a VoIP provider. Some service providers may offer plans that include free calls to other subscribers on their network and charge flat rates for other VoIP calls based on a fixed number of calling minutes. You most likely will pay additional fees when you call long distance using VoIP. While this sounds a lot like regular telephone service, it is less expensive than traditional voice communications, starting with the fact that you will no longer need to pay for extras on your monthly phone bill. [2]

New Words and Phrases

  hardware   [ˈhaːdwɛə]   n. [计算机] 硬件；五金器具；军事装备
  available   [əˈveiləbl]   adj. 可用的；可获得的；有时间的
  convert   [kənˈvəːt]   vt. 使转变；使改变信仰；改建
  recipient   [riˈsipiənt]   n. 接受者；容器   adj. 接受的

subscriber　[səbˈskraibə]　*n*. 用户；捐助者
cable modem　光缆调制解调器
Web browsers　浏览器
ISP（Internet Service Provider）　网络服务提供者
VoIP（Voice over Internet Protocol）　互联网协议电话，IP 电话
voice traffic　未来语音流量；电话业务，语音信息量，语音通信量
phone jack　听筒插口；听筒塞孔

Notes

［1］ In its simplest form, PC-to-PC Internet telephony can be as easy as hooking up a microphone to your computer and sending your voice through a cable modem to a person who has Internet telephony software that is compatible with yours.

译文：作为网络电话最简单的一种形式，计算机-计算机网络通话很简单，只要在计算机上接一个传声器，就可把声音通过调制解调器发送给与你有相匹配的网络电话软件的人。

说明：这是一个复合句。Internet telephony can be as easy as... 是主句的主语和谓语；hooking up... and sending... 是并列的动名词短语作介词 as 的宾语，句中 who 引导的定语从句修饰 person, that 引导的定语从句修饰 software；hook up 表示"接洽妥当；连接；以钩钩住"。又如：

And every Apple mobile device sold can only be activated by hooking up to the gate. 每部售出的苹果移动设备，也必须通过连接这一门户进行激活。

［2］ While this sounds a lot like regular telephone service, it is less expensive than traditional voice communications, starting with the fact that you will no longer need to pay for extras on your monthly phone bill.

译文：尽管这些听起来与普通电话差不多，但要比普通电话便宜得多，因为你不必额外再付电话月租费。

说明：句中 while 引导让步状语从句，表示"尽管；虽然"，但 while 还可表示"当……时候"引导时间状语从句，例如：Who will present his show while he's away? 他不在时，将由谁主持播放他的节目？

Exercises

Ⅰ. **Fill in the missing words according to the text.**

1. Internet telephony consists of a combination of ＿＿＿＿ and ＿＿＿＿ that enables you to use the Internet as the transmission medium for telephone calls.

2. For users who have ＿＿＿＿, or ＿＿＿＿ Internet access, Internet telephony software essentially provides free telephone calls anywhere in the world.

3. You most likely will pay ＿＿＿＿ when you call long distance using VoIP.

Ⅱ. **Translate the following sentences into Chinese.**

1. This basic form of Internet telephony is not without its problems.

2. Some, such as CoolTalk and NetMeeting, come bundled with popular Web browsers.

3. VoIP is another Internet-based communications method which is growing in popularity.

4. Much like finding an Internet service provider (ISP) for your Internet connection, you will need to use a VoIP provider.

Unit 9　Global Positioning System (GPS)

Text

GPS is the Global Positioning System. GPS uses satellite technology to enable a terrestrial terminal to determine its position on the Earth in latitude and longitude.[1]

GPS receivers do this by measuring the signals from three or more satellites simultaneously and determining their position using the timing of these signals (seen in Fig. 9-1).

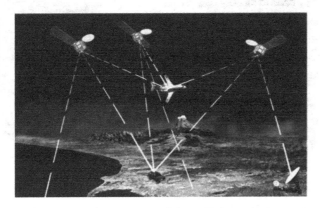

Fig. 9-1　Global Positioning System

GPS operates using trilateration. Trilateration is the process of determining the position of an unknown point by measuring the lengths of the sides of an imaginary triangle between the unknown point and two or more known points.[2]

In the GPS system, the two known points are provided by two GPS satellites. These satellites constantly transmit an identifying signal. The GPS receiver measures the distance to each GPS satellite by measuring the time each signal took to travel between the GPS satellite and the GPS receiver.

The GPS system is divided into three segments:
- The Space Segment
- The Control Segment
- The User Segment

GPS uses twenty-one operational satellites, with an additional three satellites in orbit as redundant backup. GPS uses NAVSTAR satellites manufactured by Rockwell International.[3] Each NAVSTAR satellite is approximately 5 meters wide (with solar panels extended) and weighs approximately 900kg.

GPS satellites orbit the earth at an altitude of approximately 20,200km.

Each GPS satellite has an orbital period of 11 hours and 58 minutes. This means that each GPS satellite orbits the Earth twice each day.

These twenty-four satellites orbit in six orbital planes, or paths. This means that four GPS satellites operate in each orbital plane. Each of these six orbital planes is spaced sixty degrees apart. All

55

of these orbital planes are inclined fifty-five degrees from the Equator.

In order for GPS tracking to work, it is necessary to have both access to the Global Positioning System and have a GPS receiver (seen in Fig. 9-2). [4] The GPS receiver is able to receive signals that are transmitted by GPS satellites orbiting overhead. Once these satellite transmissions are received by the GPS receiver, location and other information such as speed and direction can be calculated.

Fig. 9-2　GPS Receiver

The receiver contains a mathematical model to account for these influences, and the satellites also broadcast some related information which helps the receiver in estimating the correct speed of propagation. [5] Certain delay sources, such as the ionosphere, affect the speed of radio waves based on their frequencies; dual frequency receivers can actually measure the effects on the signals.

In order to measure the time delay between satellite and receiver, the satellite sends a repeating 1023 bit long pseudo random sequence; the receiver constructs an identical sequence and shifts it until the two sequences match. [6]

Different satellites use different sequences, which lets them all broadcast on the same frequencies while still allowing receivers to distinguish between satellites. This is an application of Code Division Multiple Access (CDMA).

There are two frequencies in use: 1575.42 MHz (referred to as L1), and 1227.60 MHz (L2). The L1 signal carries a publicly usable coarse-acquisition (C/A) code as well as an encrypted P (Y) code. The L2 signal usually carries only the P (Y) code.

New Words and Phrases

positioning　　[pə'ziʃəniŋ]　　*n.* 定位；配置，布置　　*v.* 定位；放置
terrestrial　　[tə'restrɪəl]　　*adj.* 地球的，地上的　　*n.* 陆地生物；地球上的人
terminal　　['tɜːmɪn(ə)l]　　*n.* 末端；终点；终端机　　*adj.* 末端的；终点的
latitude　　['lætɪtjuːd]　　*n.* 纬度；界限；活动范围
longitude　　['lɒn(d)ʒɪtjuːd]　　*n.* [地理] 经度；经线
simultaneously　　[ˌsɪml'teɪnɪəslɪ]　　*adv.* 同时地；

trilateration　［ˌtrailætəˈreiʃən］　n. ［测］三边测量（术）
imaginary　［ɪˈmædʒɪn(ə)rɪ］　adj. 虚构的，假想的；想象的；虚数的
triangle　［ˈtraɪæŋg(ə)l］　n. 三角形；三角形之物；三角关系；三人一组
transmit　［trænzˈmɪt］　vt. 传输；传播；发射；传达；vi. 传输；发射信号
propagation　［ˌprɒpəˈgeɪʃən］　n. 传播；繁殖；增殖
manufacture　［mænjuˈfæktʃə］　n. 制造；产品 vt. 制造；加工；捏造 vi. 制造
redundant　［rɪˈdʌnd(ə)nt］　adj. 多余的
segment　［ˈsegm(ə)nt］　n. 段；部分
orbit　［ˈɔːbɪt］　n. & v 轨道，常轨，绕轨道而行
equator　［ɪˈkweɪtə］　n. 赤道
ionosphere　［aɪˈɒnəsfɪə］　n. 电离层
pseudo　［ˈsjuːdəʊ］　adj. 假的，冒充的 n. 伪君子；假冒的人
identical　［aɪˈdentɪk(ə)l］　adj. 同一的；完全相同的 n. 完全相同的事物
sequence　［ˈsiːkw(ə)ns］　n. 序列；顺序；续发事件
frequency　［ˈfriːkw(ə)nsɪ］　n. 频率
multiple　［ˈmʌltɪpl］　adj. 多样的，多重的，许多的 n. 倍数；［电］并联
acquisition　［ˌækwɪˈzɪʃ(ə)n］　n. 获得物，获得，收购
access　［ˈækses］　n. 通路，进入，使用权 vt. 使用；存取；接近
encrypt　［enˈkrɪpt］　v. 加密，将……译成密码
enable to　使能够
be divided into　被分成
to account for　解释，说明（原因等）；对……负责；占百分之

Technicalterms

Global Positioning System（GPS）　全球定位系统
NAVSTAR　一种罗克韦尔公司制造的导航卫星
Code Division Multiple Access（CDMA）　码分多址

Notes

［1］ GPS uses satellite technology to enable a terrestrial terminal to determine its position on the Earth in latitude and longitude.

译文：GPS 通过使用卫星技术使地面终端确定它在地球上的经度和纬度位置。

说明：本句是简单句，属主动宾结构，to enable a... longitude 是不定式短语作目的状语，在该短语中 to determine... longitude 是宾语 terrestrial terminal 的补足语；GPS 是 Global Positioning System 的缩写，已经特指美国开发的全球定位系统，所以课文后续段落中 GPS system 中的 system 一词并不是多余的。

to enable... to 使……能够

例如：The shell has to be slightly porous to enable oxygen to pass in. 外壳不得不有些细小的孔以便能使氧气通过。

［2］ Trilateration is the process of determining the position of an unknown point by measuring the lengths of the sides of an imaginary triangle between the unknown point and two or more known points.

译文：三边测量法是指在未知点两侧设定两个或者两个以上的已知点，与未知点构成虚拟三角形，通过测量未知点到已知点的距离来确定未知点位置的一种测量方法。

说明：本句也是简单句，属主系表结构，the process 在句中作表语；of determining... known points 为介词加动名词作定语修饰 process，这个定语结构中又多层嵌套三个 of 引导的定语，其中 by measuring... 为方式状语，修饰 determining，该状语中还嵌套着两个 of 引导的定语结构修饰 the lengths，而 between... points 是介词短语作后置定语修饰 triangle。

［3］ GPS uses NAVSTAR satellites manufactured by Rockwell International.

译文：GPS 系统使用罗克韦尔国际公司制造的导航卫星。

说明：NAVSTAR satellites 是"导航卫星"的意思，NAV 为 navigation 的缩写；manufactured by... 是过去分词短语修饰 satellites。在英语中有几个词都有"制造"的意思，但使用的场合有些区别，学习时应该多加比较以便区分。

请比较：manufacture industry 制造业；steel fabrication 钣金制造；made in China 中国制造。

［4］ In order for GPS tracking to work, it is necessary to have both access to the Global Positioning System and have a GPS receiver.

译文：为了 GPS 进行跟踪，要能够访问全球定位系统，而且一个全球定位系统接收机也是必不可少的。

说明：In order for GPS tracking to work 为目的状语，it 作形式主语，真正的主语为不定式短语 to have both access... a GPS receiver，is necessary 是系表结构作谓语。

［5］ The receiver contains a mathematical model to account for these influences, and the satellites also broadcast some related information which helps the receiver in estimating the correct speed of propagation.

译文：接收机包含有一个数学模型来计算这些信息的影响，卫星也传播一些相关的信息以帮助接收机正确地估算传播速度。

说明：这是一个并列复合句。第一个分句是简单句；and 连接的第二个分句是复合句，主句在前，是主动宾结构；which 引导的定语从句修饰 information。to account for 在这里是负责计算的意思。propagation 是"传播"的意思。注意：在英文中尽量避免重复，在相近的语句中经常使用几个不同的词来表达近似或相同的意思，在本课中，作者使用了 transmit，broadcast，transmissions，propagation 4 个动词和名词词语来表达"传播"的意思。

［6］ The receiver constructs an identical sequence and shifts it until the two sequences match.

译文：接收机建立一个完全相同的序列，并移动它直到两个序列匹配。

说明：本句是复合句，until 引导的时间状语从句，修饰主句中的第二个谓语动词 shifts。shifts 在这里表示数据位移。

Exercises

I. **Answer the following questions according to the text.**

1. What is Global Positioning System?

2. To determine its position, what is the minimum quantity of satellites for a GPS receiver to measure their signals simultaneously?

3. What are the three segments of a GPS system?

4. Besides position or location, what kind of other information can be calculated by a GPS receiver after received satellites transmissions?

II. Match the following phrases in column A with column B.

Column A
1. positioning
2. terminal
3. trilateration
4. propagation
5. orbit
6. ionosphere
7. sequence
8. frequency

Column B
a. 频率
b. 定位
c. 传播，繁殖
d. 三角测量法
e. 终端
f. 环绕
g. 电离层
h. 次序

III. Fill in the blanks with the proper word. Change the form if necessary.

| deployment | navigation | why | during |
| justification | while | drive | base |

In the 1970s, the ground-based OMEGA navigation system, _____ on phase comparison of signal transmission from pairs of stations, became the first worldwide radio _____ system. Limitations of these systems _____ the need for a more universal navigation solution with greater accuracy.

_____ there were wide needs for accurate navigation in military and civilian sectors, almost none of those were seen as _____ for the billions of dollars it would cost in research, development, _____, and operation for a constellation of navigation satellites. _____ the Cold War arms race, the nuclear threat to the existence of the United States was the one need that did justify this cost in the view of the United States Congress. This deterrent effect is _____ GPS was funded. It is also the reason for the ultra secrecy at that time.

IV. Translation.

1. The Global Positioning System is a global navigation satellite system that provides location and time information in all weather conditions, anywhere on or near the Earth where there is an unobstructed line of sight to four or more GPS satellites.

2. The GPS project was launched in the United States in 1973 to overcome the limitations of previous navigation systems.

3. The Russian Global Navigation Satellite System was developed contemporaneously with GPS, but suffered from incomplete coverage of the globe until the mid-2000s.

4. The design of GPS is based partly on similar ground-based radio-navigation systems developed in the early 1940s and used by the British Royal Navy during World War II.

Translating Skills

被动语态的译法

英语中被动语态的使用范围极为广泛，尤其是在科技英语中，被动语态几乎随处可见，凡是在不必、不愿说出或不知道动作的执行者的情况下均可使用被动语态，因此，掌握被动语态的翻译方法极为重要。在汉语中也有被动语态，通常通过"把"或"被"等词体现出来，但它的使用范围远远小于英语中被动语态的使用范围，因此英语中的被动语态在很多情况下都翻译成主动结构。英语原文中的被动结构一般采取下面两种译法。

1. 翻译成汉语的主动句

原文的被动结构翻译成汉语的主动结构又可以进一步分为几种不同的情况。

1）原文中的主语在译文中仍作主语。在采用此方法时往往在译文中使用"加以""经过""用……来"等词体现原文中的被动含义。例如：

Other questions will be discussed briefly. 其他问题将简单地加以讨论。

In other words, mineral substances which are found on earth must be extracted by digging, boring holes, artificial explosions, or similar operations which make them available to us. 换言之，矿物就是存在于地球上，但须经过挖掘、钻孔、人工爆破或类似作业才能获得的物质。

Current flow is represented by the letter symbol I. 电流用字母 I 表示。

2）将原文中的主语翻译为宾语，同时增补泛指性的词语（如人们、大家等）作主语。例如：

It could be argued that the radio performs this service as well, but on television everything is much more living, much more real. 可能有人会指出无线电广播同样也能做到这一点，但还是电视屏幕上的节目要生动、真实得多。

3）将原文中的 by、in、for 等作状语的介词短语的宾语翻译成译文的主语，在此情况下，原文中的主语一般被翻译成宾语。例如：

A right kind of fuel is needed for an atomic reactor. 原子反应堆需要一种合适的燃料。

Special hardware (the phone adapter) is required only for the sender. 只有电话的主叫方需要特殊的硬件（电话适配器）。

4）翻译成汉语的无主句。例如：

When the information is to be conveyed over any distance a communication system is usually required. 当信息跨越一段距离传送时就需要一个通信系统。

New source of energy must be found, and this will take time... 一定要找到新能源，但这需要时间……

5）翻译成带表语的主动句。例如：

One ohm is defined as that amount of resistance that will limit the current in a conductor is one ampere when the voltage applied to the conductor is one volt. 1Ω 的定义是：$1V$ 的电压施加在导体上产生了 $1A$ 的电流时对应的导体的电阻值。

2. 译成汉语的被动语态

英语中的许多被动句可以翻译成汉语的被动句，常用"被""给""遭""挨""为……所""使""由……""受到"等表示。例如：

These signals are produced by colliding stars or nuclear reactions in outer space. 这些信号是由外层空间的星球碰撞或者核反应产生的。

When you take up the telephone, your voice can be sent to others by electricity. 当你拿起电话时，你的声音可以通过电传送到别人耳里。

The experiment must be done withthe help of our teacher. 这项实验必须在教师帮助下进行。

Over the years, tools and technology themselves as a sourcc of fundamental innovation have largely been ignored by historians and philosophers of science. 工具和技术本身可以作为根本性创新的源泉，多年来在很大程度上被科学史学家和科学思想家们忽视了。

Reading

Videoconferencing

Videoconferencing is a conference between two or more participants at different sites by using computer networks to transmit audio and video data. [1] Each participant has a video camera, microphone and speakers connected on his or her computer. As the two participants speak to one another, their voices are carried over the network and delivered to the other's speakers, and whatever images appear in front of the video camera appear in a window on the other participant's monitor.

In order for videoconferencing to work, the conference participants must use the same client or compatible software. Many freeware and shareware videoconferencing tools are available online for download, and the most Web cameras also come bundled with videoconferencing software. Many newer videoconferencing packages can also be integrated with public IM clients for multipoint conferencing and collaboration.

In recent years, videoconferencing has become a popular form of distance communication in classrooms, allowing for a cost efficient way to provide distance learning, guest speakers, and multi-school collaboration projects. [2] Many feel that videoconferencing provides a visual connection and interaction that cannot be achieved with standard IM or E-mail communications.

New Words and Phrases

videoconference ['vɪdɪoˌkɑnfərəns] n. 视频会议
participant [pɑː'tɪsɪp(ə)nt] n. 参与者；关系者 adj. 参与的；有关系的
monitor ['mɒnɪtə] n. 监视器
compatible [kəm'pætɪb(ə)l] adj. 兼容的；能共处的；可并立的
freeware ['friːwɛr] n. 免费软件；自由软件
shareware ['ʃɛrwɛr] n. 共享软件（等于 freeware）
interaction [ɪntər'ækʃ(ə)n] n. 互动；n. 相互作用；[数] 交互作用
collaboration [kəlæbə'reɪʃn] n. 合作；勾结；通敌
IM (Instant Messaging) 即时通信
bundled with 与……捆绑

Notes

[1] Videoconferencing is a conference between two or more participants at different sites by using computer networks to transmit audio and video data.

译文：视频会议是指在两个或两个以上的不同地点的人通过用计算机网络传递声音和视频数据在一起开会。

说明：此句是一个稍显复杂的简单句，主系表结构。句中 between... sites 是介词短语作后置定语，修饰 conference；by using... data. 是介词短语作方式状语。

[2] In recent years, videoconferencing has become a popular form of distance communication in classrooms, allowing for a cost efficient way to provide distance learning, guest speakers, and multi-school collaboration projects.

译文：最近几年，视频会议已成为教室中远距离通信的一种流行方式，是用来实现远程学习、嘉宾演讲和多校合作项目的一种很有效的节约成本的方式。

说明：此句是简单句，但句子较长。介词 of + 名词词组构成的定语修饰 form；allowing for 是现在分词短语作伴随情况的状语。allow for 表示"考虑到，虑及"，例如：

Ideally, you want to allow for this in the original system specification. 在理想的情况下，您需要在原先的系统规格中考虑到这一点。

Exercises

Ⅰ. Decide whether the following statements are True (T) or False (F) according to the text.

1. Videoconferencing is a conference between two participants at different sites by using computer networks to transmit audio and video data.

2. In order for videoconferencing to work, the conference participantscan use different client or compatible software.

3. Videoconferencing is more powerful than standard IM.

4. most Web cameras come bundled with videoconferencing software.

Ⅱ. Translation.

1. 电视会议是一个利用多种通信技术的会议解决方案，通过同时双向视频和音频的传输来实现两地或多的沟通。

2. 电视会议与视频电话的不同之处在于：这个设计是用来实现会议或者为多地服务而不是为个体。

3. 编解码器通常用于视频会议和流媒体解决方案。

Unit 10　4G Network Technology

Text

  4G is the fourth generation of wireless communications currently being developed for high speed broadband mobile capabilities. It is characterized by higher speed of data transfer and improved quality of sound. Although not yet defined by the ITU (International Telecommunications Union), the industry identifies the following as 4G technologies:
- WIMAX (Worldwide Interoperability for Microwave Access);
- 3GPP LTE (3rd Generation Partnership Project Long Term Evolution);
- UMB (Ultra Mobile Broadband);
- Flash-OFDM (Fast Low-latency Access with Seamless Handoff Orthogonal Frequency Division Multiplexing).

  The 4G technology is being developed to meet QoS (Quality of Service) and rate requirements that involve prioritization of network traffic to ensure good quality of services. [1] These mechanisms are essential to accommodate applications that utilize large bandwidth such as the following (as shown in Fig. 10-1): [2] Wireless Broadband Internet Access, MMS (Multimedia Messaging Service), Video Chat, Mobile Television, HDTV (High Definition TV), DVB (Digital Video Broadcasting), Real Time Audio, High Speed Data Transfer.

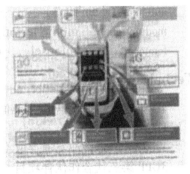

Fig. 10-1　4G Network

  The goal set by ITU for data rates of WIMAX and LTE is to achieve 100Mbit/s when the user is moving with high speed relative to the base station, and 1Gbit/s for fixed positions. [3]

  The industry moves towards expansion of the number of 4G compatible devices. It is set to find its way to tens of different mobile devices not restricted to 4G phones or laptops, such as Video Camera, Gaming Devices, Vending Machines and Refrigerators.

  The trend is to provide wireless internet access to every portable device that could supply and incorporate the 4G embedded modules. The 4G technology could not only provide internet broadband connectivity but also a high level of security that is beneficial to devices that incorporate financial transactions such as vending machines and billing devices. [4]

  The following key features can be observed in all suggested 4G technologies:
- MIMO: to attain ultra-high spectral efficiency by means of spatial processing including multi-antenna and multi-user MIMO.
- Frequency-domain-equalization, for example, multi-carrier modulation (OFDM) in the downlink or single-carrier.

Frequency-domain-equalization (SC-FDE) in the uplink: To exploit the frequency selective channel property without complex equalization.
- Turbo principle error-correcting codes: to minimize the required SNR at the reception side.

63

- Channel-dependent scheduling: to use the time-varying channel.
- Link adaptation: adaptive modulation and error-correcting codes.
- Mobile-IP utilized for mobility.
- IP-based femtocells (home nodes connected to fixed Internet broadband infrastructure).

New Words and Phrases

capability ['keipə'biləti] n. 才能，能力；性能，容量
access ['ækses] vt. 使用；存取；接近 n. 进入；使用权；通路
evolution [ˌiːvəˈluːʃ(ə)n; 'ev-] n. 演变；进化论；进展
interoperability [ˈintərˌɒpərəˈbiləti] n. 互通性，互操作性
latency ['leitənsi] n. 潜伏时间；延迟时间
seamless ['siːmlis] adj. 无缝的；不停顿的
handoff ['hændɔːf] n. 切换；传送；（美国橄榄球）手递手传球
traffic ['træfik] n. 交通；通信量 vt. 用……作交换；在……通行 vi. 交易
mechanism ['mekənizəm] n. 机制；原理，途径；进程；机械装置；技巧
embed [imˈbed] vt. 栽种；使嵌入，使插入；使深留脑中
spectral ['spektrəl] adj. [光] 光谱的
antenna [ænˈtenə] n. [电信] 天线；[动] 触角，[昆] 触须
modulation [ˌmɒdjʊˈleɪʃən] n. [电子] 调制；调整
femtocell ['femtəusel] n. 家庭基站

Technical Terms

4G 第四代移动电话通信标准
ITU 国际电信联盟
WiMAX 全球微波互通存取
3GPP 第三代合作伙伴项目
LTE 长期演进技术
QoS 服务质量技术
UMB 超移动宽带
OFDM 正交频分复用技术
MMS 多媒体短信服务
HDTV 高清电视
DVB 数字视频广播
MIMO (Multi-input Multi-output) 多输入多输出技术
SC-FDE 单载波频域均衡
SNR (signal-to-noise ratio) 信噪比
embedded module 嵌入模块
spectral efficiency 频谱效率
adaptive modulation 自适应调制

Notes

[1] The 4G technology is being developed to meet QoS (Quality of Service) and rate require-

ments that involve prioritization of network traffic to ensure good quality of services.

译文：正在开发的 4G 技术可以满足服务质量和网络流量优先化的速率要求，以保证良好的服务质量。

说明：本句是复合句，that involve... services 是限定性定语从句修饰 requirements.

［2］ These mechanisms are essential to accommodate applications that utilize large bandwidth such as the following.

译文：这些机制对于大宽带的调节应用来说是十分重要的。

说明：to accommodate applications 不定式短语作状语修饰形容词 essential；that utilize large bandwidth such as the following 为定语从句修饰 applications；accommodate 具有"调节，使适应；使适合；使符合一致；调和；通融，给……提供方便"的意思，又如：

She tried to accommodate her way of life to his. 她试图使自己的生活方式与他的生活方式相适应。

She will accommodate me with the use of her car. 她会把她的汽车供我使用。

［3］ The goal set by ITU for data rates of WIMAX and LTE is to achieve l00Mbit/s when the user is moving with high speed relative to the base station, and 1Gbit/s for fixed positions.

译文：由 ITU 为 WIMAX 和 LTE 设定的目标是：当用户相对于基站高速移动时，数据传输速率达到 100Mbit/s；当相对于基站位置固定时，数据传输速率达到 1Gbit/s。

说明：set by... 是过去分词短语作后置定语修饰 goal；to achieve 100Mbit/s 为不定式短语作表语；when 引导时间状语从句来修饰动词 achieve。

set by 在句中可理解为"设定"，此外还有"搁在一旁；抛开，撇开；留出"的意思，例如：

It's time to set our differences by and work together for a common purpose. 该是我们抛开分歧，为一个共同的目的而协力工作的时候了。

［4］ The 4G technology could not only provide internet broadband connectivity but also a high level of security that is beneficial to devices that incorporate financial transactions such as vending machines and billing devices.

译文：4G 技术不仅可以提供宽带互联网连接而且具有更高级别的安全性，有利于像自动售货机和计费装置等包含金融交易的设备的安全运行。

说明：本句是复合句，句中 that is beneficial... devices 为定语从句修饰 security；定语从句中 that incorporate... devices 为嵌套的定语从句来修饰句子中第一个 devices。

Exercises

I. Answer the following questions according to the text.

1. What are the two characters of 4G technology according to the text?
2. What is the goal set by ITU for data rates of WIMAX and LTE?
3. What kind of 4G devices that incorporate financial transactions are mentioned in the text?

II. Decide whether the following are true (T) or False (F) according to the text.

1. 4G is characterized by high speed of data transfer and improved quality of sound.
2. The goal set by ITU for data rates of WIMAX and LTE is to achieve 10Mbit/s when the user is moving with high speed relative to the base station, and 100Mbit/s for fixed positions.
3. 4G is set to find its way to mobile devices only restricted to 4G phones or laptops.
4. 4G technology will provide the same level of security as 3G.

Ⅲ. Choose the best answer for the following sentences.

1. 4G is _____ by higher speed of data transfer and improved quality of sound.
   A. feature   B. characterized   C. property

2. These mechanisms are _____ to accommodate applications that utilize large bandwidth.
   A. importance   B. necessity   C. essential

3. The goal _____ ITU for data rates of WIMAX and LTE is to achieve 100Mbps when the user is moving with high speed relative to the base station.
   A. set by   B. set on   C. set up

4. The trend is to provide wireless internet access _____ every portable device that could supply and incorporate the 4G embedded modules.
   A. with   B. to   C. by

5. The following key features can be _____ in all suggested 4G technologies.
   A. seem   B. find   C. observed

Ⅳ. Translate the following sentences into English.

1. 4G 是具有高速移动宽带能力的第四代无线通信技术。
2. 正在开发的 4G 技术可以满足 QoS 和网络流量优先化的速率要求从而保证良好的服务质量。
3. 目前的趋势是给每一个提供和纳入 4G 嵌入式模块的便携式装置提供无线互联网接入。
4. 4G 的特点是更高的数据传输速率和声音质量。

Translating Skills

非谓语动词 V-ing 的用法

1. 名词特性

V-ing 作动名词时具有名词的特征，因而在句中用作主语、表语、宾语和定语。

（1）作主语、表语

Learning how to make a clean sketch is necessary for every student. 学会如何作清洁的草图对每个学生都是必要的。（主语）

Their work is *producing PCB* (printed curcuit board). 他们的工作是生产印制电路板。（表语）

（2）作宾语

This machine needs *oiling*. 这台机器需要加油。（作及物动词的宾语）

There are different ways of *producing electricity*. 有各种不同产生电的方法。（作介词的宾语）

（3）作定语

V-ing 作定语时往往表示被修饰词的用途和作用，如：

Many newer *videoconferencing* packages can also be integrated with public IM clients for multi-point conferencing and collaboration。许多较新款的视频会议（软件）包还能与即时通信客户端软件集成，用于多点开会和合作。

2. 形容词或副词特性

V-ing 充当非谓语动词作现在分词时具有形容词或副词的特征，可在句中充当定语、表语、宾补及状语。

（1）作定语

1）现在分词作定语时表示该动作正在进行。单个现在分词作定语通常放在被修饰词的前面，现在分词短语作定语则放在被修饰词之后。

Does a *flying* bird have any weight? 飞鸟有重量吗？

The factory *producing drawing instruments* was built last year. 这个生产绘图仪器的工厂是在去年建成的。

2）现在分词与所修饰的名词有逻辑上的主谓关系。

Packet switching is similar to a jigsaw puzzle—the image that the puzzle represents is divided into pieces at the *manufacturing* factory and put into a plastic bag. 分组交换就像智力拼图玩具——玩具的组件在工厂被拆开放入塑料袋中。（现在分词相当于定语从句 which are manufacturing）

（2）作表语

V-ing 作表语说明主语具备的性质，可以用 very、so 或 much 等修饰。

The book is very *interesting*. 这本书很有趣。

（3）作宾语补足语

V-ing 作宾语补足语表明宾语正在进行的动作。

Electricity sets machines *running*. 电使机器运转。

（4）作状语

1）现在分词作状语时，一般用逗号与句子的其他部分分开，其位置可在基本部分的前面或后面，可以表示时间、原因、条件、结果等。

Working in the factory, we learned much from the workers. 在工厂工作期间，我们向工人学到了许多东西。（时间）

Being very short, the waves of a laser beam cannot be seen directly by man. 由于激光的波长很短，人们不能直接看见。（原因）

Compressing a gas, we can change it into a liquid. 将气体压缩，我们能将它变为液体。（条件或方式）

A substance can combine with another substance, *forming a new one*. 一种物质和另一种物质相化合能形成一种新的物质。（结果）

The teacher is standing there, *reading English*. 教师站在那里读英语。（伴随）

2）现在分词作状语，其逻辑主语一般与句子的主语相同。如果句中主语不是它的逻辑主语，V-ing 也可以带自己的主语，构成独立主格结构，如：

Fuel burning, we get a large amount of heat. 燃料燃烧时，我们就得到大量的热。（独立主格结构，表示时间）

Reading

I-Mode

While a multi-industry consortium of telecom vendors and computer companies was busy hammering out an open standard using the most advanced version of HTML available, other developments were going on in Japan. There, a Japanese woman, Mari Matsunaga, invented a different approach to the wireless Web called i-mode (information-mode). She convinced the wireless subsidiary of the former Japanese telephone monopoly that her approach was right, and in February 1999, NTT DoCoMo (literally: Japanese Telephone and Telegraph Company everywhere you go) launched the service in Japan.[1] Within three years it had over 35 million Japanese subscribers, who could access over 40,000 special i-mode Web sites. It also had most of the world's telecom companies drooling over its financial success, especially in light of the fact that WAP appeared to be going nowhere. Let us now take a look at what i-mode is and how it works.

The i-mode system has three major components: a new transmission system, a new handset and a new language for Web page design. The transmission system consists of two separate networks: the existing circuit-switched mobile phone network (somewhat comparable to D-AMPS), and a new packet-switched network constructed specifically for i-mode service. Voice mode uses the circuit switched network and is billed per minute of connection time. I-mode uses the packet-switched network and is always on (like ADSL or cable), so there is no billing for connect time. Instead, there is a charge for each packet sent. It is not currently possible to use both networks at once.

The handsets look like mobile phones, with the addition of a small screen. NTT DoCoMo heavily advertises i-mode devices as better mobile phones rather than wireless web terminals, even though that is precisely what they are. In fact, probably most customers are not even aware they are on the Internet. They think of their i-mode devices as mobile phones with enhanced services. In keeping with this model of i-mode being a service, the handsets are not user programmable, although they contain the equivalent of a PC in 1995 and could probably run Windows 95 or UNIX.

When the i-mode handset is switched on, the user is presented with a list of categories of the officially-approved services. There are well over 1,000 services divided into about 20 categories. Each service, which is actually a small i-mode Web site, is run by an independent company. The major categories on the official menu include e-mail, news, weather, sports, games, shopping, maps, horoscopes, entertainment, travel, regional guides, ringing tones, recipes, gambling, home banking and stock prices. The service is somewhat targeted at teenagers and people in their 20s, who tend to love electronic gadgets, especially if they come in fashionable colors. The mere fact that over 40 companies are selling ringing tones says something. The most popular application is e-mail, which allows up to 500B messages, and thus is seen as a big improvement over SMS (short message service) with its 160B messages. Games are also popular.

Current handsets have CPUs that run at about 100MHz, several megabytes of Flash ROM, perhaps 1MB of RAM, and a small built-in screen. I-mode requires the screen to be at least 72pixels × 94pixels, but some high-end devices have as many as 120 pixels × 160 pixels. Screens usually have 8 bit color, which allows 256 colors. This is not enough for photographs but is adequate for line drawings and simple cartoons. Since there is no mouse, on-screen navigation is done with the arrow keys.

New Words and Phrases

consortium [kən'sɔːiəm] n. 财团；联合；合伙
subsidiary [səb'sidiəri] adj. 辅助的，补充的；子公司
approach [ə'prəutʃ] n. 方法；途径；接近 vt. 接近；着手处理 vi. 靠近
subscriber [səb'skraibə(r)] n. 订户；
monopoly [mə'nɑpəli] n. 垄断；专利权
drool [druːl] v. 流口水；说昏话
comparable ['kɑmpərəbl] adj. 可比较的，比得上的
D-AMPS Digital-Advanced Mobile Phone System 时分多址联接方式

Programmable   [ˌprəʊˈɡræməbl]   adj. 可设计的，可编程的
UNIX   n. UNIX 操作系统（一种多用户的计算机操作系统）
switch on   接通，开启
electronic gadgets   数码产品
hammer out   打造

Notes

[1] She convinced the wireless subsidiary of the former Japanese telephone monopoly that her approach was right, and in February 1999, NTT DoCoMo (literally: Japanese Telephone and Telegraph Company everywhere you go) launched the service in Japan.

译文：她让以前日本电信垄断运营商的无线子公司相信她的方法是正确的。1999 年 2 月，NTT DoCoMo（字面意思是：伴你行日本电话电报公司）在日本启动了 I-mode 服务。

说明：本句是并列复合句。第一个分句是复合句，主句是主语＋动词＋间宾＋直宾结构，但直接宾语是由 that 引导的宾语从句充当的。第二个分句是简单句，两分句之间由并列连词 and 连接。

Exercises

Ⅰ. Answer the following questions according to the text.
1. Who invented a different approach to the wireless Web called I-mode?
2. In which year the I-mode service was launched in Japan?
3. What are the three major components of I-mode?
4. What is the up limit of e-mail application of I-mode?
5. Why there is an on-screen navigation key on I-mode device?

Ⅱ. Translate the following phrases and expressions into Chinese.
1. information mode   2. transmission system   3. fashionable colors
4. high-end device   5. in light of

Unit 11 Smartphone

Text

You probably hear the term " Smartphone" tossed around a lot. But if you've ever wondered exactly what a Smartphone is, well, you're not alone. What is a Smartphone different than a cell phone, and what makes it so smart?[1]

In a nutshell, a Smartphone is a device that lets you make telephone calls, but also adds in features that, in the past, you would have found only on a personal digital assistant or a computer, such as the ability to send and receive E-mail and edit Office documents. [2]

However, to really understand what a Smartphone is (and is not), we should start with a history lesson. In the beginning, there were cell phones and personal digital assistants (or PDAs). Cell phones were used for making calls —and not much else —while PDAs, like the Palm Pilot, were used as personal, portable organizers. A PDA could store your contact information and a to-do list, and could SYNC with your computer.

PDAs gained wireless connectivity and were able to send and receive E-mail. Cell phones, meanwhile, gained messaging capabilities, too. PDAs then added cellular phone features, while cell phones added more PDA-like (and even computer-like) features. The result was the Smartphone.

Unlike many traditional cell phones, Smartphone allow individual users to install, configure and run applications of their choosing (as shown in Fig. 11-1). A Smartphone offers the ability to conform the device to your particular way of doing things. Most standard cell phone software offers only limited choices for re-configuration, forcing you to adapt to the way it's set up. On a standard phone, whether or not you like the built-in calendar application, you are stuck with it except for a few minor tweaks. If that phone were a Smartphone, you could install any compatible calendar application you like. For example, a Smartphone will enable you to do more. It may allow you to create and edit Microsoft documents or at least view the files. It may allow you to download APPs as well.

Smart phones are run by using Operating Systems just like a computer does, some of the more popular operating systems or OS are the Android, Windows, Symbian and Blackberry OS. [3]

The Android OS was released in 2008 and it is considered an open source platform that was supported by Google. The features

Fig. 11-1 Smartphone

it included were Maps, Calendar, Gmail and a fully functioning HTML web browser. The extremely popular Smartphone is the iPhone from Apple. It was first introduced to the world in 2007 and it priced at a high $499. The iPhone was the first phone to have a large touch screen that was designed for direct finger input.

Since cell phones and PDAs are the most common handheld devices today, a Smartphone is usually either a phone with added PDA capabilities or a PDA with added phone capabilities (as shown in Fig. 11-2). Here's a list of some of the things Smartphones can do:
- Send and receive mobile phone calls – some Smartphones are also WIFI capable;
- Personal Information Management (PIM) including notes, calendar and to-do list;
- Communication with laptop or desktop computers;
- Data synchronization with applications like Microsoft Outlook and Apple's iCal calendar programs;
- E-mail;
- Instant messaging;
- Applications such as word processing programs or video games;
- Play audio and video files in some standard formats.

While most cell phones include some software, even the most basic models these days have address books or some sort of contact manager.

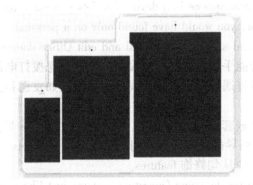

Fig. 11-2 Smartphone with Added PDA Capabilities

New Words and Phrases

connectivity   [kɒnek'tɪvɪtɪ]   n. 连接性；连通性可连接性
cellular   ['seljulə]   adj. 细胞的；多孔的；由细胞组成的   n. 移动电话
application   [ˌæplɪ'keɪʃ(ə)n]   n. 应用；申请；应用程序；敷用
configure   [kən'fɪgə]   vt. 配置，设置，按特定形式装配，安装
stuck   [stʌk]   v. 刺（stick 的过去式）   adj. 动不了的；被卡住的
tweak   [twiːk]   vt. 稍稍调整（机器、系统等）
platform   ['plætfɔːm]   n. 平台；坛；讲台平台
laptop   ['læptɒp]   n. 膝上型轻便式计算机，笔记本式计算机
pilot   ['paɪlət]   adj. 试验性的；导向的；驾驶员的；辅助的
tossed around   翻来覆去
in a nutshell   简而言之
adapt to   使适应于

TechnicalTerms

cell phone（=cellular phone） 移动电话
PDA（personal digital assistants） 个人数字助手
SYNC（synchronous communication） 同步通信
WIFI　Wireless Fidelity 的缩写，基于 IEEE 802.11b 标准的无线局域网
HTML　Hyper Text Markup Language 的缩写，超文本标记语言
Palm Pilot　Palm 公司开发的一款掌上式计算机

Notes

[1] How is a Smartphone different than a cell phone, and what makes it so smart?

译文：智能手机和一个普通手机相比到底有什么不同，是什么使它这么智能呢？

说明：此句是并列句，两个分句都是简单句。different than 意思是"与……不同"，在美国英语中一般情况下等同于 different from，且使用的频率比 different from 高。如：Seeing is different than looking. 看见和看到是不同的。

[2] In a nutshell, a Smartphone is a device that lets you make telephone calls, but also adds in features that, in the past, you would have found only on a personal digital assistant or a computer, such as the ability to send and receive E-mail and edit Office documents.

译文：简而言之，智能手机是这样一个设备，它能让您拨打电话，并且增加了其他一些您以前在掌上式计算机或普通计算机里才能发现的功能，如发送和接收电子邮件、编辑微软 Office 文档。

说明：这是一个具有嵌套结构的复合句。主句在前，device 后边是 that 引导定语从句，修饰 device；第一个 that 引导的从句包含并列的谓语结构，即 lets you make... 与 adds in 并列；第二个 that 引导的定语从句修饰 features。

[3] Smart phones are run by using Operating Systems just like a computer does, some of the more popular operating systems or OS are the Android, Windows, Symbian and Blackberry OS.

译文：智能手机是通过类似于计算机使用的操作系统来运行的，一些比较流行的操作系统有安卓、视窗、塞班和黑莓等。

说明：Android 原意为"机器人偶，仿真机器人"。谷歌公司开发的一种主要用于便携设备，以 Linux 为基础的开放源代码操作系统以此命名。中国大陆一般以"安卓"作为译名。

Symbian 中文译为"塞班"，是基于 EPOC32 的一种手机操作系统，在早期的智能手机上使用率很高。

Blackberry OS 是加拿大通信公司 RIM 开发的基于 Java 的手机操作系统。中国大陆很多客户习惯以"Blackberry"（黑莓）这个词指代 RIM 的公司名称。

Exercises

I. Answer the following questions according to the passage.

1. What is a Smartphone?
2. According to the text, what is the basic function of a cell phone or cellular phone?
3. Can you mention some of the functions of Smartphone listed in the text?
4. When the Android OS was released by Google?

Ⅱ. Match the following phrases in column A with column B.

Column A
1. connectivity
2. application
3. configure
4. platform
5. pilot
6. HTML
7. Operating Systems
8. Wireless Fidelity

Column B
a. 应用
b. 连接性
c. 平台
d. 配置
e. 超文本标记语言
f. 无线局域网
g. 试验性的
h. 操作系统

Ⅲ. Fill in the blanks with the proper word. Change the form if necessary.

| refer | as well as | make | able |
| apps | prototype | through | demonstrate |

The first mobile phone to incorporate PDA features was anIBM prototype developed in 1992 and _____ that year at the COMDEX computer industry trade show. The _____ demonstrated PDA features _____ other visionary apps like maps, stocks and news incorporated with a cellular phone. A refined version of the product was marketed to consumers in 1994 by BellSouth under the name Simon Personal Communicator. The Simon was the first cellular device that can be properly _____ to as a "smartphone", although it was not called that in 1994. In addition to its ability to _____ and receive cellular phone calls, Simon was _____ to send and receive faxes and emails and included several other _____ like address book, calendar, appointment scheduler, calculator, world time clock, and note pad _____ its touch screen display.

Ⅳ. Translate the following sentences into Chinese.

1. In 1999, the Japanese firmNTT DoCoMo released the first smart phone to achieve mass adoption within a country.

2. Symbian was the most popular smart phone OS in Europe during the middle to late 2000s. .

3. In 2013, it was reported thatfoldable OLED smart phones could be as much as a decade away because of the anticipated cost of producing them is dropping.

4. iPhone introduced numerous design concepts that have been adopted by modern smart phone platforms.

Translating Skills

非谓语动词 V-ed 和 to V 的用法

1. 非谓语动词 V-ed 的用法

过去分词 V-ed 可以在句子中作表语、定语、状语和补足语，一般具有"被动"和"完成"的意义。

（1）作表语

I was *excited* at the news. 听到这个消息我很兴奋。

（2）作定语

The energy *released* is very great. 释放出来的能量很大。

（3）作状语

过去分词在句子中作状语表示原因、时间、条件、伴随状况等，表示被动或完成的动作。

Encouraged by the teacher, we worked still harder than ever. 被老师鼓励后，我们比以前更努力地学习。（原因状语，相当于 As we were encouraged by the teacher）

Given s and t, we can find *v*. 已知距离和时间，我们就能求出速度。（条件状语）

This done, he went on with the next. 这件工作做完以后，他就进行下一件。（独立分词短语作时间状语）

（4）作补足语

He found the tube *broken*. 他发现管子破了。

2. 非谓语动词 to V 的用法

动词不定式 to V 可在句中作主语、宾语、表语、定语、状语、补足语。

（1）作主语

To finish the job in two days is impossible. 要在两天之内完成这项工作是不可能的。

（2）作宾语

I want *to know* something about laser beams. 我想知道一些关于激光的知识。

（3）作表语

Their work is *to make* airplanes of this kind. 他们的工作就是制造这种飞机。

（4）作定语

Compressed air has the ability *to do work*. 压缩空气具有做功的能力。

（5）作状语

This machine is easy *to operate*. 这台机器容易操作。

（6）作补足语

Scientists found the waves of a laser beam *to be very short*. 科学家们发现激光波长很短。

Reading

E-mail & Instant Messaging

If you use the internet, then you probably use Internet-based communications to contact family, friends or co-workers. From sending an instant message to a friend, to E-mailing co-worker, to conducting videoconferences, the Internet offers a number of ways to communicate.

The advantages of Internet-based communications are many. Since you are already paying for an Internet account (or your employer is), you can save money on phone calls by sending someone an instant message or by using VoIP instead of standard local telephone services. Of course, no technology is without a downside and Internet-based communications has plenty, such as viruses, privacy issues and spam. Like all technologies (and especially technology tied to the Internet), the way we can communicate online is constantly evolving. Here we'll take a look at some of the most popular forms of Internet based communications.

1. E-mail

Short for electronic mail, E-mail is the transmission of messages over communications networks. Most mainframes, minicomputers and computer networks have an E-mail system. Some E-mail systems are confined to a single computer system or network, but others have gateways to other computer systems, enabling you to send electronic mail anywhere in the world.[1]

Using an E-mail client (software such as Microsoft Outlook or Eudora), you can compose an E-mail message and send it to another person anywhere, as long as you know the recipient E-mail address. All online services and Internet Services Providers (ISPs) offer E-mail, and support gateways so that you can exchange E-mail with users of other systems. Usually, it takes only a few seconds for an E-mail to arrive at its destination. This is a particular effective way to communicate with a group because you can broadcast a message or document to everyone in the group at once.

One of the biggest black clouds hanging over E-mail is spam. Though definitions vary, spam can be considered any electronics junk mail (generally E-mail advertising for some product) that is sent out to thousands, if not millions, of people. Often spam perpetrates the spread of E-mail Trojans and viruses. For this reason, it's important to use an updated antivirus program, which will scan your incoming and outgoing E-mail for viruses.

2. Instant Messaging

One of the fastest-growing forms of Internet communications is instant messaging or IM. Think of IM as a text-based computer conference between two or more people, an IM communications service enables you to create a kind of private chat room with another individual in order to communicate in real-time over the Internet.[2] Typically, the IM system alerts you whenever somebody on your buddy or contact list is online. You can then initiate a chat session with that particular individual.

While IM is used by millions of Internet users to contact family and friends, it's also growing in popularity in the business world. Employees of a company can have instant access to managers and Co-workers in different offices and can eliminate the need to place phone calls when information is required immediately. Overall, IM can save time for employees and help decrease the amount of money a business spends on communications.

New Words and Phrases

account [əˈkaʊnt] *n.* 账目，理由；账户 *vi.* 解释
downside [ˈdaʊnsaɪd] *n.* 下降趋势；负面，消极面 *adj.* 底侧的
spam [spæm] *n.* 垃圾邮件
evolve [ɪˈvɒlv] *vt.* 发展，进化；推断出 *vi.* 发展，进化；逐步形成
mainframe [ˈmeɪnfreɪm] *n.* 大型主机
confine [kənˈfaɪn] *vt.* 限制；禁闭 *n.* 界限，边界
perpetrate [ˈpɜːpɪtreɪt] *vt.* 犯（罪）；做（恶）
trojan [ˈtrəʊdʒən] *n.* 特洛伊木马，木马程式
antivirus [ˈæntivaɪrəs] *n.* 抗病毒素；反病毒程序
buddy [ˈbʌdɪ] *n.* 伙伴，好朋友，[美口]密友 *vi.* 做好朋友，交朋友
popularity [ˌpɒpjʊˈlærətɪ] *n.* 普及，流行；名气；受大众欢迎
VOIP *abbr.* (Voice over Internet Phone) 网络语音电话业务

Notes

[1] Some E-mail systems are confined to a single computer system or network, but others have gateways to other computer systems, enabling you to send electronic mail anywhere in the world.
译文：有些电子邮件系统是局限于单个计算机系统或网络中的，但有些可通过网关送到

连接到其他计算机系统，可以把电子邮件送到世界的任何一个地方。

说明：本句是并列句，由 but 连接两个分句。两个分句都是简单句，但第二个分句包含有 enabling... 作伴随情况的状语。

［2］Think of IM as a text-based computer conference between two or more people, an IM communications service enables you to create a kind of private chat room with another individual in order to communicate in real-time over the Internet.

译文：可把 IM 看成是在两人或更多人之间的计算机文字会议，IM 通信服务器可以创建个人聊天室，使你与另一人在网上实时聊天。

说明：这是用逗号连接的并列句，第一分句是祈使句，第二分句是比较复杂的简单句，其复合谓语是 enables... to create，宾语是 a kind... chat room，之后是不定式短语作状语。

Exercises

Ⅰ. Decide whether the following are true (T) or False (F) according to the text.

1. Internet-based communications has plenty of downside including viruses, privacy issues and spam.

2. Often spam perpetrates the spread of E-mail Trojans and viruses.

3. IM is used by millions of Internet users to contact family and friends and not used in business world.

Ⅱ. Translate the following sentences into English.

1. 及时信息（IM）是一种互联网在线聊天业务，它能够通过互联网提供及时文字信息传递。

2. 电子邮件在 1993 年左右被大多数人称为 email 或者 e-mail，但是这种拼写的变体也被使用。

3. Skype 是一种互联网语音服务，它提供计算机对计算机的免费呼叫或者计算机对电话的折扣价呼叫，这种服务在 2003 年发布。

Unit 12 5G Technology

Text

5G Technology stands for 5th generation mobile technology. 5G mobile technology has changed the means to use cell phones within very high bandwidth. User never experienced ever before such a high value technology. Nowadays mobile users have much awareness of the cell phone (mobile) technology. The 5G technologies include all type of advanced features which makes 5G mobile technology most powerful and in huge demand in the near future.

The giganticarray of innovative technology being built into new cell phones is stunning. 5G technology which is on hand held phone offering more power and features than at least 1000 lunar modules. [1] A user can also hook their 5G technology cell phone with their Laptop to get broadband internet access. 5G technology including camera, MP3 recording, video player, large phone memory, dialling speed, audio player and much more you never imagine. For children rocking fun Bluetooth technology and Piconet have become a reality in market.

What 5G Technology Offers

5G technology is going to be a new mobile revolution in mobile market. Through 5G technology now you can use worldwide cellular phones and this technology also strike the China mobile market and a user being proficient can get access to Germany phone as a local phone. With the coming out of cell phone alike to PDA now your whole office is in your finger tips or in your phone. 5G technology has extraordinary data capabilities and has ability to tie together unrestricted call volumes and infinite data broadcast within latest mobile operating system. 5G technology has a bright future because it can handle best technologies and offer priceless handset to their customers. May be in coming days 5G technology takes over the world market. 5G Technologies have an extraordinary capability to support software and consultancy. The router and switch technology used in 5G network providing high connectivity. [2] The 5G technology distributes internet access to nodes within the building and can be deployed with union of wired or wireless network connections. The current trend of 5G technology has a glowing future.

Features of 5G Technology

- 5G technology offer high resolution for crazy cell phone user and bi-directional large bandwidth shaping.
- The advanced billing interfaces of 5G technology makes it more attractive and effective.
- 5G technology also providing subscriber supervision tools for fast action.
- The high quality services of 5G technology based on policy to avoid error.
- 5G technology is providing large broadcasting of data in Gigabit which supporting almost 65,000 connections.
- 5G technology offers transporter class gateway with unparalleled consistency.
- The traffic statistics by 5G technology makes it more accurate.
- Through remote management offered by 5G technology a user can get better and fast solution.
- The remote diagnostics also a great feature of 5G technology.

- The 5G technology is providing up to 25 Mbps connectivity speed.
- The 5G technology also supports virtual private network.
- The new 5G technology will take all delivery service out of business prospect.
- The uploading and downloading speed of 5G technology touches the peak.
- The 5G technology network offers enhanced and available connectivity just about the world.
- A new revolution of 5G technology is about to begin because 5G technology is going to give tough competition to normal computers and laptops whose marketplace value will be affected. [3] There are lots of improvements from 1G, 2G, 3G, and 4G to 5G in the world of telecommunications. The new coming 5G technology is available in the market in affordable rates, high peak future and much reliability than its preceding technologies.

New Words and Phrases

bandwidth ['bændwɪdθ] n. 频带宽度
array [ə'reɪ] n. 数组,阵列 vt. 排列,部署;打扮
lunar ['luːnə] adj. 月亮的,月球的;阴历的
stunning ['stʌnɪŋ] adj. 极好的;使人晕倒的;震耳欲聋的
dial ['daɪəl] n. 转盘;刻度盘;钟面 vi. 拨号
proficient [prə'fɪʃnt] adj. 熟练的,精通的 n. 精通;专家,能手
handset ['hæn(d)set] n. 手机,电话扬声器
consultancy [kən'sʌlt(ə)nsɪ] n. 咨询公司;顾问工作
unparalleled [ʌn'pærəleld] adj. 无比的;无双的;空前未有的
router ['ruːtə(r)] n. 路由器
gateway ['geɪtweɪ] n. 门;网关;方法;通道;途径
deploy [dɪ'plɔɪ] vt. 配置;展开;使疏开 vi. 部署;展开 n. 部署
subscriber [səb'skraɪbə(r)] n. 订户;签署者;捐献者
diagnostic [daɪəg'nɒstɪk] adj. 诊断的;特征的 n. 诊断法;诊断结论
virtual ['vɜːtʃuəl] adj. 虚拟的;事实上的(但未在名义上或正式获承认)
stands for 代表;支持;象征;担任……的候选人
have awareness of 了解到,意识到
piconet 微微网
billing interface 计费接口

Notes

[1] 5G technology which is on hand held phone offering more power and features than at least 1000 lunar modules.

译文:手持电话中的5G技术将提供至少1000个月球登月舱才能够提供的功能。

说明:这是一个含有定语从句的复合句,which引导定语从句,修饰technology;主句谓语动词省略了"is"是为了避免重复,这是科技英语写作中对未来表示肯定的一种用法;Lunar module 指"月球登月舱"。

[2] The router and switch technology used in 5G network providing high connectivity。

译文:在5G技术中使用的路由器和转换技术将提供很高的连通性。

说明:这是一个简单句。used in 5G network 是过去分词短语作后置定语修饰 the router

and switch technology。5G 网络还没有真正商业化，但句中却使用现在分词形式 providing 且省略了系动词 be，这仍然是科技英语写作中对未来表示肯定的一种用法，而且在本文章的其他地方也可以找到类似的应用。

[3] A new revolution of 5G technology is about to begin because 5G technology going to give tough competition to normal computers and laptops whose marketplace value will be affected.

译文：5G 技术将使普通计算机和笔记本式计算机在竞争中生存很艰难，它们的市场价值将受到影响，而一个新的 5G 技术革命将要发生。

说明：这是一个双重嵌套的复合句。主句在前，because 引导原因状语从句，而从句本身又包含了 whose 引导的定语从句修饰 computers and laptops。句中 going to 前面可认为省略了助动词 is。在科技英语中省略助动词 be 和不定式省略 to 的情况比较常见，这种情况下一般不会对语义产生误解。作者对语法的严谨性有时并不十分在意，而更在意语气的表达。

Exercises

Ⅰ. Answer the following questions according to the passage.

1. What is the most outstanding characteristic of 5G according to paragraph one?
2. According to the text what technology will provide high connectivity in 5G?
3. How fast is the large databroadcast speed in 5G and how many connections it will support?
4. What extend the 5G will give competition to normal computers and laptops?

Ⅱ. Match the following phrases in column A with column B.

Column A	Column B
1. bandwidth	a. 路由器
2. array	b. 频谱带宽
3. handset	c. 网关
4. router	d. 手机，传声器
5. gateway	e. 诊断
6. diagnostic	f. 计费接口
7. billing interface	g. 订户
8. subscriber	h. 排列

Ⅲ. Translate the following sentences into Chinese.

1. Mobile generations typically refer to non - backward-compatible cellular standards following requirements stated by ITU-R, such as IMT-2000 for 3G and IMT-Advanced for 4G.
2. 5G technology offer high resolution for crazy cell phone user and bi-directional large bandwidth shaping.
3. The advanced billing interfaces of 5G technology makes it more attractive and effective.
4. 5G technology is providing large broadcasting of data in Gigabit which supporting almost 65,000 connections.

Translating Skills

定语从句的翻译

定语从句一直是科技类英语阅读中经常遇到却又难以把握的内容。定语从句一般位于所修饰的名词（先行词）后，紧跟先行词，或与先行词有一定的间隔。定语从句究竟修饰文中的哪个词，一要靠语言知识理解，二要根据上下文进行逻辑判断。

定语从句可以分为限制性定语从句和非限制性定语从句两种。限制性定语从句对修饰的先行词起限制作用，在意义上与先行词密不可分，是主句意义中不可缺少的部分，一般不可用逗号与主句分开；而非限制性定语从句仅对先行词在原因、让步方面进行解释，或对时间、条件、结果等进行补充说明，关系代词不能省略，从句与主句用逗号分开，引导词有 which（不是 that）. when. who 等。例如：

A transformer is composed of some coils *that* are coupled together by magnetic induction. （that 引导限制性定语从句。）译为：变压器由磁感应耦合线圈组成。

ISDN is a new service offered by many telephone companies *that* provides fast, high-capacity digital transmission of voice, data, still images and full motion video over the worldwide telephone network. （that 引导限制性定语从句。）译为：综合业务数据网（ISDN）是由电话公司提供的一项新型业务，这项业务可以在世界范围的电话网上进行语音、数据、静止图像和全动态视频的快速、大容量数字传输。

When the pointer stops moving, the reading given by the ohmmeter is the insulation resistance, *which* is normally high if the capacitor is in good condition. （which 引导非限制性定语从句。）译为：指针停止移动时的读数就是电容器的绝缘电阻值，假如电容器正常工作，绝缘电阻值通常很高。

First of all, let's introduce you to some of the circuits that are commonly found in radio-frequency systems and *with which* you may not be familiar. （介词 + which 引导定语从句。）类似的句型还有介词 + whom 等引导定语译为：首先让我们向你介绍射频系统中常见的一些电路，或许你并不熟悉它们。

The analog world is full of relative numbers, tradeoffs, and approximations, *all of which* depend heavily on the basic semiconductor properties. （代词 + 介词 + which 引导定语从句，句中的 all 指代前面的数值等。译为：模拟电子技术领域充满了相对数、折中和近似值，所有这些在很大程度上取决于半导体的基本特性。

Direct current is an electric current *the charges of which* move in one direction only. 名词 + 介词 + which 引导定语从句。）译为：直流电是电荷只向一个方向流动的电流。

注意：

1) 在限制性定语从句中，引导词指代先行词，定语从句的引导词在从句中充当某个句子成分，当其充当宾语时可将其省略。例如：

The GPS receiver compares the time a signal was transmitted by a satellite with the time it was received. GPS 接收机把卫星发送信号的时间和接收信号的时间相比较。

2) 在非限制性定语从句中不能使用 that 作为关系代词，而只能用 who、whom、whose 等指人，用 which 指物或事。例如：

Microwaves, *which* have a higher frequency than ordinary radio waves, are used routinely in sending thousands of telephone calls and television programs across long distance. （本句中用 which 引导的非限制性定语从句解释原因。）译为：微波比一般无线电波信号的频率更高，它通常被用来进行远距离大量的长途电话和电视节目的传送。

3) 在非限制性定语从句中，定语从句的引导词不仅可以指代先行词，有时还指代前面的整个句子。例如：

There have been some promising discoveries in battery research, *which* may hasten the development of a practical battery-powered car. 在电池研究领域的一些有前景的发现，可能会加速

实用电动汽车的开发。

Reading

5G Cellular Systems Overview

As the different generations of cellular telecommunications have evolved, each one has brought its own improvements. [1] The same will be true of 5G technology.

First generation, 1G: These phones were analogue and were the first mobile or cellular phones to be used. Although revolutionary in their time they offered very low levels of spectrum efficiency and security.

Second generation, 2G: These were based around digital technology and offered much better spectrum efficiency, security and new features such as text messages and low data rate communications.

Third generation, 3G: The aim of this technology was to provide high speed data. The original technology was enhanced to allow data up to 14 Mbit/s and more.

Fourth generation, 4G: This was an all-IP based technology capable of providing data rates up to 1 Gbit/s.

Any new 5th generation, 5G cellular technology needs to provide significant gains over previous systems to provide an adequate business case for mobile operators to invest in any new system.

Facilities that might be seen with 5G technology include far better levels of connectivity and coverage. The term World Wide Wireless Web, or WWWW is being coined for this.

For 5G technology to be able to achieve this, new methods of connecting will be required as one of the main drawbacks with previous generations is lack of coverage, dropped calls and low performance at cell edges. 5G technology will need to address this.

New Words and Phrases

generation　　[dʒenəˈreɪʃ(ə)n]　　n. 一代人；代（约30年），时代
analogue　　[ˈænəlɒg]　　n. 相似物；adj. 模拟计算机的；类似的，相似物的
spectrum　　[ˈspektrəm]　　n. 光谱；频谱；范围
cellular　　[ˈseljʊlə]　　adj. 细胞的；多孔的　n. 移动电话；单元
digital　　[ˈdɪdʒɪtl]　　adj. 数字的；手指的　n. 数字；键；
enhance　　[ɪnˈhɑːns]　　vt. 提高；加强；增加
connectivity　　[kɒnekˈtɪvɪti]　　n. 连通性
coverage　　[ˈkʌvərɪdʒ]　　n. 覆盖，覆盖范围；新闻报道
drawback　　[ˈdrɔːbæk]　　n. 缺点，不利条件；退税
address　　[əˈdres]　　vt. 演说；向……致辞；提出；处理　n. 地址；演讲；致辞
IP　　[ˈaɪˈpi]　　abbr. 互联网协议（Internet Protocol）

Notes

[1] As the different generations of cellular telecommunications have evolved, each one has brought its own improvements.

译文：由于不同时代的蜂窝通信技术都进行了演变，每一代技术都有它的改进和提升。

说明：此句是复合句。As 引导原因状语从句，后面是主句；句中的 one 指代前面的 generation；evolve 表示"发展，进化，演变"；involve 表示"包含，牵涉，潜心于"。例如：

This would most likely involve an abstract aspect with an abstract pointcut. 这很有可能涉及一个带有抽象切入点的抽象方面。

But the universe as we observe it seems to evolve that way. 但宇宙貌似正如我们所观测的那样在不断地进行演变。

Exercises

Ⅰ. **Decide whether the following statements are True（T）or False（F）according to the text.**

1. The first generation cellular technology was based on analog and provide very low levels of spectrum and security.

2. From the second generation of cellular technology and on, they began to based on digital technology.

3. 3G can provide high speed data rate up to 600 Mbit/s and more.

4. 4G was an all-IP based technology capable of providing data rates up to 1Gbit/s.

Ⅱ. **Translate the following sentences into English.**

1. 日益增长的移动互联网接入、云服务和大数据分析，使得任何人在任何地点都能利用"大智慧"———一个全球连接和共享的知识基础。

2. 5G 将提供智慧城市建设的重要基础设施，这也将推动移动网络的表现和能力需要到极致。

3. 5G 的成功只能建立在整个 ICT 生态系统成功的基础上。

Chapter III Computer Technology

Unit 13 Computer Systems

Text

A computer system is made up of a number of different sub-component systems, which together allow the system to perform calculations and complicated tasks. [1] A computer system could run payrolls, control an engine in a car, fly an airplane or allow a user to play games and balance their checkbook.

Computer systems do vary in size, cost, and power, depending upon the task that they are required to perform. [2] In this course, we are primarily concerned about personal computer systems suitable for small business or home use.

Base Unit

The base unit holds the computer's motherboard, on which the computer's memory storage area [commonly referred to as RAM (random access memory)] and CPU (central processing unit) are located. RAM holds both programs and data, the larger the RAM size, the more complicated the programs that can be run, and the more data that can be processed. RAM is measured in Megabytes (one GB = 1 billion characters). Typical size for today's personal computer systems is 500GB.

The floppy drive and hard drive are storage devices that are used to keep permanent copies of programs or data. The floppy drive supports a removable media disk, which the user can take away and use on another computer system. The hard drive is considered a non-removable media disk because it is permanently fixed inside the base unit. Floppy drives support up to about 1.44MB of data, while fixed disks support 500GB upwards (512000MB or 500billion characters).

The CPU is the device that actually runs all the programs and processes the data. It's like the motor of the car, it does all the work and makes things happen. The PSU (power supply unit) is also located in the base unit and provides power to the memory, CPU and other devices.

Keyboard

Keyboards allow users to enter commands and data into the computer system. An example of a command might be to run a checkbook program or dial up a remote computer.

Monitor

Monitors are devices that allow the computer to display information back to the user. This might be in either a text or graphical display. Monitors come in various sizes, 20″, 22″, 24″, and so on. The larger the monitor, the more expensive it is, and the larger the image displayed on the screen is [3].

Monitors have a number of important features. Screen resolution refers to the number of dots in the X and Y co-ordinates (1400×900, or 1920×1080), and refresh rate specifies the number of times per second that the image is drawn on the screen (60Hz means 60 times per second). You can learn more about monitors here. Higher screen resolutions like 1400×900 require a large monitor size like 21″ (otherwise it looks so small when viewed on a 14″ monitor), and also require a higher refresh rate like 72Hz, in order to prevent the image on the screen flickering.

Mouse

A mouse is an inverted trackball device that has a number of selection buttons associated with it. The hand moves it across a flat surface, and its position is displayed on the screen. As the mouse is moved across the surface, its position displayed on the screen is updated to reflect its movement. The buttons are used to select items and make choices on the screen.

The mouse can significantly reduce user input by moving away from typing commands at the keyboard to clicking on buttons or other items displayed on the screen.

Printer

A printer allows the user to print out on paper a copy of the screen or the data that is being processed by the computer. Printers are available in both color and black and white. Color printers are slower and more expensive than black and white printers. In addition, the technology used to print the information on the paper varies upon the type of printer.

Modem

A modem is a device that allows the user to connect their computer system to another computer system. A modem attaches to a telephone line, and dials up another computer via the telephone.

The modem conveys the computer signals so that they work over the telephone circuits used by the telephone companies. A modem can be an internal device that is located inside the base unit, or an external device that attaches to the base unit via a cable.

New Words and Phrases

calculation [ˌkælkjuːleiʃən] n. 计算；估计；计算的结果
complicated ['kɔmplikeitid] adj. 难懂的，复杂的
payroll ['peirəul] n. 工资单；发放的工资总额
checkbook ['tʃekbuk] n. 支票簿；核算
permanent ['pəːmənənt] adj. 永久的，永恒的 n. [口] 烫发
display [disˈplei] vt. 显示；表现；陈列 n. 显示；炫耀
resolution [ˌrezəˈluːʃən] n. 分辨率；决议；解决
specify ['spesifai] vt. 指定；详细说明；规定
flicker ['flikə] vi. 闪烁；摇曳；颤动 vt. 使闪烁 n. 闪烁；闪光
personal computer systems 个人计算机系统
base unit 主机
referred to as 被称为
floppy drive 软盘驱动器
refresh rate 刷新率

Notes

[1] A computer system is made up of a number of different sub-component systems, which together allow the system to perform calculations and complicated tasks.

译文：计算机系统由许多不同的子元器件组成，这些元器件组合起来可进行计算或完成

复杂的任务。

说明：本句是复合句，which 引导非限制性定语从句，修饰 sub-component systems；be made up of 意思是"由……组成，由……构成"。

［2］Computer systems do vary in size, cost, and power, depending upon the task that they are required to perform.

译文：计算机设备的大小、成本及功能差异很大，主要取决于它要完成什么工作。

说明：句中 depending upon the task that they are required to perform 是现在分词短语做伴随状语，that 引导定语从句，修饰 task。depend upon 意思是"取决于，依靠，依赖"。

［3］The larger the monitor, the more expensive it is, and the larger the image displayed on the screen is.

译文：显示器越大，价格越贵，屏幕显示的影像也就越大。

说明：句中 displayed on the screen 是过去分词短语做后置定语，修饰 image。英语中"the + 比较级……，the + 比较级……"结构用于表示随着前事物的变化，后事物呈相应的变化，译为"越……，越……"。

Exercises

Ⅰ. Answer the following questions according to the passage.
1. What sub-component systems does a typical personal computer system consist of?
2. What is the central processing unit?
3. What does screen resolution refer to?
4. What device is used for communication between computers?

Ⅱ. Translate the following phrases into Chinese or English.
1. base unit　　　　　　＿＿＿＿＿＿＿＿＿＿
2. 随机存储器　　　　　＿＿＿＿＿＿＿＿＿＿
3. CPU　　　　　　　　＿＿＿＿＿＿＿＿＿＿
4. 软盘驱动器　　　　　＿＿＿＿＿＿＿＿＿＿
5. screen resolution　　　＿＿＿＿＿＿＿＿＿＿
6. 刷新率　　　　　　　＿＿＿＿＿＿＿＿＿＿

Ⅲ. Fill in the blanks with the proper word. Change the form if necessary.

| type | individual | keyboard | monitor |
| inexpensive | browse | addition | storage |

PC (personal computer) is a small, relatively ＿＿＿＿＿ general purpose computer designed for an ＿＿＿＿＿ user. "General purpose" means that you can do many different things with a PC. You can use it to ＿＿＿＿＿ documents, send e-mail, ＿＿＿＿＿ the web and play games. In ＿＿＿＿＿ to the microprocessor, a personal computer has a ＿＿＿＿＿ for entering data, a ＿＿＿＿＿ for displaying information, and a ＿＿＿＿＿ device for saving data.

Ⅳ. Translation.
1. User interface provides a way for you to communicate and interact with the computer.
2. Virtual memory is the space on a hard disk used to temporarily store data and swap it in and out of RAM as needed.
3. A minicomputer is a multiprocessing system capable of supporting from 4 to about 200 users

simultaneously.

Translating Skills

And 引导的句型的译法

And 作为连词，用来连接词、短语和句子，其基本意义相当于汉语的"和""与""并且"。但在实际翻译的过程中，特别是在连接两个句子时，它的译法很多，表达的意义可能相差甚远。如果不考虑 and 前后成分之间的逻辑关系，只用几种基本译法生硬套用，难免造成理解上的失误，甚至把整个句子意思搞错。

1) 表示原因，例如：

Laser is widely used for developing many new kinds of weapons, and it penetrates almost everything. 激光广泛用于制造各种新式武器，因为它的穿透力很强。

2) 表示结果，例如：

But since a digital signal is made up of a string of simple pulses, noise stands out and easily removed. 但由于数字信号由一组简单脉冲组成，杂音明显，因而容易排除。

In 1945 a new type of aeroplane engine was invented, it was much lighter and powerful than earlier engines, and enabled war planes to fly faster and higher than ever. 1945 年发明了一种新型的飞机发动机，它比早期的发动机要轻得多，功率也要大得多，因此采用这种发动机的军用飞机比以往任何时候都飞得更快、更高。

3) 表示目的，例如：

It was later shown that the results of this work were by no means the ultimate, and further work has been put in hand and to provide closer control and more consistent operation in this area. 后来发现，这项研究工作的结果绝非已作定论，而且进一步的研究工作已开始，以便在这方面提供较严密的控制和持续的操作。

4) 表示承接，例如：

In many ways, computer is more superior than human brain, and human can rule it. 在许多方面计算机超过人脑，而人却可以控制它。

5) 表示对照，例如：

Motion is absolute, and stagnation is relative. 运动是绝对的，而静止是相对的。

6) 表示递进，例如：

The electronic brain calculates a thousand times quicker, and more accurately than is possible for the human being. 计算机的运算速度比人所能达到的速度要快 1000 倍，而且更加准确。

7) 表示转折，例如：

There will always be some things that are wrong, and that is nothing to be afraid of. 错误在所难免，但并不可怕。

8) 表示条件，例如：

Even if a programmer had endless patience, knowledge and foresight, storing every relevant detail in a computer, the machine's information would be useless, and the programmer knew little how to instruct it in what human beings refer to as commonsense reasoning. 即使一个编程员很有耐心、知识和远见，把每一个有关细节都存入计算机，如果他不懂得按人类常识推理去对计算机下达指令，机器里的信息也还是没有用途的。

9) 表示结果，例如：

Operators found that the water level was too low so they turned on two additional main coolant pumps, and too much cold water flowing into the system caused the steam to condense, further destabilizing the reactor. 操作人员发现冷却水的水位过低,就启动了另外两台主冷却泵,结果过量的冷却水进入系统使蒸气冷凝,反应堆因而更不稳定。

Reading

BIOS (Basic Input/Output System)

The BIOS is a special software that interfaces the major hardware components of your computer with the operating system.[1] It is usually stored on a Flash memory chip on the motherboard, but sometimes the chip is another type of ROM (Read Only Memory).

The BIOS software has a number of different roles, but its most important role is to load the operating system. When you turn on your computer and the microprocessor tries to execute its first instruction, it has to get that instruction from somewhere. It cannot get it from the operating system because the operating system is located on a hard disk, and the microprocessor cannot get to it without some instructions that tell it how. The BIOS provides those instructions. Some of the other common tasks that the BIOS performs include:

- A power-on self-test (POST) for all of the different hardware components in the system to make sure everything is working properly.
- Activating other BIOS chips on different cards installed in the computer. For example, SCSI and graphics cards often have their own BIOS chips.
- Providing a set of low-level routines that the operating system uses to interface to different hardware devices. It is these routines that give the BIOS its name.[2] They manage things like the keyboard, the screen, and the serial and parallel ports, especially when the computer is booting.

The first thing the BIOS does is to check the information stored in a tiny (64 bytes) amount of RAM located on a CMOS chip. The CMOS setup provides detailed information particular to your system and can be altered as your system changes. The BIOS uses this information to modify or supplement its default programming as needed.

New Words and Phrases

software ['sɔftweə] n. 软件
interface ['intəfeis] v. (使通过界面或接口)接合,连接;[计]使联系
microprocessor [ˌmaikrəu'prəusesə] n. [计]微处理器
execute ['eksikjuːt] vt. 实行;执行;处死
activate ['æktiveit] vt. 刺激;使活动
graphics ['græfiks] n. [测]制图学;制图法;图表算法
routine [ruː'tiːn] n. 常规,程序 adj. 日常的;例行的
manage ['mænidʒ] vi. 处理 vt. 管理,控制,操纵
alter ['ɔːltə] vt. 改变,更改
supplement ['sʌplimənt] vt. 增补,补充

Notes

[1] The BIOS is a special software that interfaces the major hardware components of your computer with the operating system.

译文：BIOS 是一种特殊的软件，是主要硬件与操作系统的接口软件。

说明：本句是复合句，句中 that 引导定语从句，修饰 software。Interface... with... 意思是"与……相连接，与……配合"。例如：

First, we need to find a way to interface with a business process from PHP. 首先，我们需要找到一种方法来为从 PHP 使用业务流程提供接口。

[2] It is these routines that give the BIOS its name.

译文：就是这些低级程序调用各种设备。

说明：此句是一个强调句。强调句句型为 It is/ was + 被强调部分（通常是主语、宾语或状语） + that/ who（当强调主语且主语指人） + 其他部分，构成强调句的 it 本身没有词义；强调句中的连接词一般只用 that、who，即使在强调时间状语和地点状语时也如此，that 也不可省略。

Exercises

Ⅰ. Decide whether the following statements are True (T) or False (F) according to the text.

1. The most important role of the BIOS software is to load the operating system.

2. The microprocessor can get instruction from the operating system because the operating system is located on a hard disk.

3. The first thing the BIOS does is to check information stored in a tiny amount of ROM located on a CMOS chip.

Ⅱ. Translate the following paragraph into Chinese.

The first thing the BIOS does isto check the information stored in a tiny (64 bytes) amount of RAM located on a CMOS chip. The CMOS setup provides detailed information particular to your system and can be altered as your system changes. The BIOS uses this information to modify or supplement its default programming as needed.

Unit 14 Computer Operating System

Text

Every desktop and laptop computer in common use today contains a microprocessor as its central processing unit. To get its work done, the microprocessor executes a set of instructions known as software. The operating system is one kind of software. [1]

The operating system provides a set of services for the applications running on your computer, and it also provides the fundamental user interface for your computer.

The purpose of an operating system is to organize and control hardware and software so that the device it lives in behaves in a flexible but predictable way.

All desktop computers have operating systems. The most common are the Windows family of operating systems developed by Microsoft, the Macintosh operating systems developed by Apple and the UNIX family of operating systems. For a desktop computer user, this means you can add a new security update, system patch, new application or often even a new operating system entirely rather than junk your computer and start again with a new one when you need to make a change.

An operating system does two things.

Firstly, it manages the hardware and software resources of the system, as various programs and input methods compete for the attention of the central processing unit (CPU) and demand memory, storage and input/output (I/O) bandwidth for their own purposes. In this capacity, the operating system plays the role of the good parent, making sure that each application gets the necessary resources while playing nicely with all the other applications, as well as husbanding the limited capacity of the system to the greatest good of all the users and applications. [2]

Secondly, it provides a stable, consistent way for applications to deal with the hardware without having to know all the details of the hardware. It is especially important if there is more than one of a particular type of computer using the operating system, or if the hardware making up the computer is ever open to change. A consistent application program interface (API) allows a software developer to write an application on one computer and have a high level of confidence that it will run on another computer of the same type, even if the amount of memory or the quantity of storage is different on the two machines. [3]

New Words and Phrases

instruction [inˈstrʌkʃ(ə)n] n. 指令，命令；指示；教导；用法说明
fundamental [fʌndəˈment(ə)l] adj. 基本的，根本的 n. 基本原则
organize [ˈɔːgənaiz] vt. 组织；使有系统化
flexible [ˈfleksibl] adj. 柔韧的，易曲的，灵活的
predictable [priˈdiktəb(ə)l] adj. 可预言的
security [siˈkjuərəti] n. 安全
update [ʌpˈdeit] vt. 更新；校正，修正 n. 现代化，更新
patch [pætʃ] n. 片，补缀，碎片 vt. 修补；解决；掩饰

compete　　［kəmˈpiːt］　　*vi.* 比赛，竞争
storage　　［ˈstɔːridʒ］　　*n.* 存储；仓库；贮藏所
bandwidth　　［ˈbændwitθ］　　*n.* ［电子］［物］带宽；［通信］频带宽度
consistent　　［kənˈsistənt］　　*adj.* 一致的，一贯的，相容的
desktop computer　　台式计算机
laptop computer　　便携式计算机
system patch　　系统补丁
application program　　应用程序

Notes

［1］ To get its work done, the microprocessor executes a set of instructions known as software.

译文：工作时，微处理器执行一系列的指令，这些指令称为软件。

说明：句中不定式短语 to get its work done 作目的状语；make \ get \ have sth. done 表示"使某事被做、请别人做某事"的意思，其中 done 是过去分词，作宾语补足语；known as software 是过去分词短语作定语，修饰 a set of instructions，相当于一个定语从句。

［2］ In this capacity, the operating system plays the role of the good parent, making sure that each application gets the necessary resources while playing nicely with all the other applications, as well as husbanding the limited capacity of the system to the greatest good of all the users and applications.

译文：这时操作系统起到一个好的管理作用，它保证各个应用程序得到必要的资源并与其他应用程序协调工作，同时面向所有的用户和应用程序，使系统的有限资源得到最充分的应用。

说明：句中 making sure... all the other applications 和 husbanding... all the users and applications 都是现在分词短语作状语，补充说明操作系统的作用；play the role of 意思是"起着……作用、担任……角色"；as well as 表示"不但……而且、和……一样"。

［3］ A consistent application program interface (API) allows a software developer to write an application on one computer and have a high level of confidence that it will run on another computer of the same type, even if the amount of memory or the quantity of storage is different on the two machines.

译文：一个一致的应用程序界面（API）使一个软件开发者在一台计算机上编写的应用程序可以在其他任何同类的计算机上运行，即使两台计算机的内存或储存量不同，也没关系。

说明：句中 that it will run on another computer of the same type, even if the amount of memory or the quantity of storage is different on the two machines 是同位语从句，用来说明 confidence 的具体含义或内容；引导词 that 在同位语从句中是连词，只起连接作用，无具体词义，that 不可省略；even if 意思是"虽然，即使"，引导让步状语从句。

Exercises

Ⅰ. Answer the following questions according to the passage.

1. What are the three most common operating systems?
2. What is the purpose of an operating system?
3. What things does an operating system do?

Ⅱ. Decide whether the following statements are True (T) or False (F) according to the text.

1. The microprocessor in the desktop and laptop computer is used as its central processing unit.
2. Not all desktop computers have operating systems.

3. As a desktop computer user, you must start again with a new computer when you need to make a change.

4. The operating system can manage both the hardware and software resources of the system.

5. The operating system makes the greatest use of all the users and applications.

Ⅲ. Translate the following phrases into Chinese or English.
1. 台式计算机
2. 笔记本式计算机
3. 软件资源
4. the UNIX family of operating systems
5. system patch
6. an application program

Ⅳ. Translate the following sentences into Chinese.
1. In order to define the control structures (e.g., tables) that the O/S needs to manage processes and resources, it must have access to configuration data during initialization.

2. The principal responsibility of the operating system is to control the execution of processes.

3. A typical UNIX system employs two Running states, to indicate whether the process is executing in user mode or kernel mode.

Translating Skills

虚拟语气的翻译

科技文章中常常提出一些设想、推理或判断，内容与事实相反，或者不大可能实现。为了同客观存在的实际相区别，要求用虚拟语气。

1. 简单虚拟句

简单虚拟句由 should/would/could/might + 动词原形构成，表示主观对客观事物的看法、愿望、请求、建议等。

The receiver should recover the baseband signal exactly. 接收机应完全恢复基带信号。

In this case, there would be no need to have security mechanisms within the network itself. 在这种情况下，网络本身不需要建立密码机制。

2. 虚拟语气用于 if 条件句

If there were no attraction between the proton and the electron, the electron would fly away from the proton in a straight line. 倘若质子与电子间不存在引力，电子就会沿直线飞离质子。（表示与现在事实相反）

The bit stream that we finally arrive at is smaller than what would have resulted had we used ASCII/UNICODE Tables. 最终我们所得到的比特流，比我们采用 ASCII/UNICODE 得到的要小。（表示与过去事实相反）

如果上述条件状语从句中省略 if，句子的主语和谓语动词部分要倒装。

If we were to travel by space rocket to the moon, we would found it quite a different place from our own planet—the earth. 若我们乘宇宙飞船到月球上去旅行，我们将会发现月球与我们自己的星球——地球是完全不一样的。（表示与将来事实相反）

3. 虚拟语气用于宾语从句

在 desire、demand、suggest、recommend、advise、decide、order 等表示"愿望、建议、

命令、要求、忠告"的动词的宾语从句中往往用"should+动词原形"或动词原形的虚拟语气。

High efficiency <u>implies</u> that circuit loss <u>be</u> minimum and the ratio of the transistor output, the parallel equivalent resistance, and its collector load resistance be maximum. 高效率意味着电路损耗应最小,而晶体管输出比率、并联等效电阻及集电极负载电阻应最大。

Communication over long distances usually <u>requires</u> that some alterations or other operations be performed on the electrical signal conveying the information in preparation for transmission. 远距离通信,通常需要对拟传输的承载信息的电信号进行某些变换或操作。

4. 虚拟语气用于主语从句

"It is (was) +形容词(或过去分词、名词) + that..."结构中的虚拟语气,其表达形式为should + 动词原形或省略should 直接用动词原形。

常用形容词有 appropriate、advisable、preferable、necessary、important、imperative、urgent、essential、vital、possible、compulsory、crucial 等。

常用过去分词有 required、demanded、desired、suggested、ordered 等。

常用名词有 advice、decision、desire、demand、order、preference、proposal、recommendation、requirement、resolution、suggestion 等。

It is <u>necessary</u> that the theoretical sections of the book <u>be</u> carefully studied and thoroughly understood. 必须仔细研究并透彻理解本书的理论部分。

It is my <u>proposal</u> that he <u>study</u> the information theory first. 我建议他应该先学信息论。

5. 虚拟语气用于状语从句

The sum of the kinetic energy and potential energy of the system is always the same, <u>provided</u> the system be not acted upon by anything outside it. 假定系统不受任何外力作用,系统的动能和势能之和就始终保持不变。(表示条件)

Audio signals are attenuated very rapidly, so we <u>would</u> have to be within a few hundred feet of the originating source in order to hear the signal. 音频信号衰减地很快,所以要在距离声源几百英尺范围以内才能听到它。(表示结果)

Reading

Windows XP

Windows XP is the next version of Microsoft Windows 2000 and Windows Millennium. Windows XP is the convergence of Windows operating systems by integrating the strengths of Windows 2000.

While maintaining the core of Windows 2000, Windows XP features a fresh new visual design. Common tasks have been consolidated and simplified, and new visual cues have been added to help you navigate your computer more easily.

Windows XP makes it possible for fast user switching in multiple users of a computer. Windows XP takes advantage of Terminal Services technology and runs each user session as a unique Terminal Services session, enabling each user's data to be entirely separated.[1] Fast user switching makes it easier for families to share a single computer.

Windows XP has new visual styles and themes that use sharp 24bit color icons and unique colors that can be easily related to specific tasks.[2] Windows XP makes it easy to keep track of your files by letting you arrange them in various groups. You can view your documents by type. You can also group files according to the last time you modified them such as today, yesterday, last week, two months ago, earlier this year or last year.

Windows XP uses webview technology helping you better manage files and the file namespace. This functionality is similar to what you see in Windows 2000 if you right-click on a file or folder. Windows XP takes this information and brings it into view directly on the desktop.

Windows XP introduces an easier-to-manage taskbar by grouping multiple instances of the same application. Windows XP makes user interface enhance productivity. The new user interface takes the windows operating system to a new level of usability, enabling you to complete tasks more easily and faster than ever before.

Windows XP features Windows Media Player 8, which brings together common digital media activities including CD and DVD playback, jukebox management and recording, audio CD creation, Internet radio playback, and media transfer to portable devices. Within Windows XP, the new "My Music" folder makes common music tasks easier to perform.

New Words and Phrases

version ['vəːʃən] n. 译文，版本，翻译
millennium [mi'leniəm] n. 千年期，千禧年；一千年，千年纪念
convergence [kən'vəːdʒəns] n. 集中，收敛
strength [streŋθ] n. 力量；力气；兵力；长处
maintain [mein'tein] vt. 维持，继续
visual ['vizjuəl] adj. 看的，视觉的，形象的
consolidate [kən'sɔlideit] v. 巩固，加强
simplify ['simplifai] vt. 单一化，简单化
navigate ['nævigeit] v. 航行，航海，航空，操纵，使通过
multiple ['mʌltipl] adj. 多样的，多重的 n. 倍数，若干 v. 成倍增加
specific [spi'sifik] adj. 特殊的，特定的；明确的；详细的
modify ['mɔdifai] vt. 更改，修改
folder ['fəuldə] n. 文件夹，折叠器，折叠机
enhance [in'hɑːns] vt. 提高，增强
productivity [ˌprɔdʌk'tivəti] n. 生产力；生产率；生产能力
usability ['juːzəbləti] n. 合用，可用；可用性
transfer [træns'fəː] n. 移动，传递，转移 vt. 转移，传递，转让
keep track of 跟踪，记录，保持联系

Notes

[1] Windows XP takes advantage of Terminal Services technology and runs each user session as a unique Terminal Services session, enabling each user's data to be entirely separated.

译文：Windows XP 利用终端服务技术的优势，把每一个用户会话作为一个独立的终端服务会话，这样确保了每一个用户的数据都是完全独立的。

说明：句中 enabling each user's data to be entirely separated 是现在分词短语作结果状语；take advantage of 意思是"利用"。

[2] Windows XP has new visual styles and themes that use sharp 24bit color icons and unique colors that can be easily related to specific tasks.

译文：Windows XP 具有新的视觉风格和主题，使用清晰的 24 位彩色图标以及容易与特定任务相联系的独特颜色。

说明：句中 that use sharp 24bit color icons 和 that can be easily related to specific tasks 都是 that 引导的限制性定语从句，分别修饰 visual styles and themes 和 unique colors；be related to 意思是"与……相联系，与……有关"。

Exercises

Ⅰ. Decide whether the following statements are True (T) or False (F) according to the text.

1. Windows XP maintains the major strengths of Windows 2000.
2. Windows XP has new visual styles and themes that use sharp 32-bit color icons.
3. WindowsXP is characterized by Windows Media Player 10.

Ⅱ. Translate the following paragraph into Chinese.

While maintaining the core of Windows 2000, Windows XP features a fresh new visual design. Common tasks have been consolidated and simplified, and new visual cues have been added to help you navigate your computer more easily.

Unit 15 Programming Language

Text

C is commonly considered to be a structured language, in an academic sense, C is an informal part of that language group. The distinguishing feature of block-structured language is the compartmentalization of code and data. This means that a language can section off and hide, from the rest of the program, all information and instructions that are necessary to perform a specific task. [1]

Functions are the building blocks of C, in which all program activity occurs. They allow you to define and code specific tasks in a program separately. After debugging a function that uses only local variables, you can rely on it to work properly in various situations without creating side effects in other parts of your program. [2] All variables that are declared in that function will be known only to that function.

In C, using blocks of code also creates program structure. A block of code is a logically connected group of program statements that can be treated as a unit. You create a block of code by placing lines of code between opening and closing curly braces. In C, every statement can be either a single statement or a block of statements. The use of code block creates readable programs with logic that is easy to follow.

C is a programmer's language. Unlike most high-level computer languages, C imposes few restrictions on what you can do with it. [3] By using C, a programmer can avoid the use of assembly code in all but most demanding situations. In fact, one reason for the invention of C was to provide an alternative to assembly language programming.

New Words and Phrases

structure ['strʌktʃə] n. 结构，构造
academic [ˌækə'demik] adj. 学院的，理论的
distinguish [dis'tiŋgwiʃ] vt. 区别，辨别
compartmentalization [ˈkɔmpɑːtˌmentəlaiˈzeiʃən] n. 区分，划分
perform [pəˈfɔːm] vt. 履行，执行 vi. 完成任务
debug [diːˈbʌg] vt. 调试 n. [计] 调试工具
variable [ˈvɛəriəbl] n. [数] 变数，变量 adj. 可变的，[数] 变量的
logically [ˈlɔdʒikəli] adv. 理论上，逻辑上
assembly [əˈsembli] n. 集合，汇编
restriction [riˈstrikʃən] n. 限制，约束
alternative [ɔːlˈtəːnətiv] n. 二中择一，可供选择的办法、事物
curly brace 花括号，大括号
a programmer's language 编程语言
assembly language 汇编语言

Notes

[1] This means that a language can section off and hide, from the rest of the program, all information and instructions that are necessary to perform a specific task.

译文：这意味着这种语言可以把完成一个指定任务所需要的信息和指令与程序的其他部分分离开，并隐藏起来。

说明：句中第一个 that 引导宾语从句；第二个 that 引导限制性定语从句，修饰 all information and instructions.

[2] After debugging a function that uses only local variables, you can rely on it to work properly in various situations without creating side effects in other parts of your program.

译文：在调试过一个只用局部变量的函数以后，这个函数可以在各种情况下正常运用，不会对程序的其他部分产生副作用。

说明：句中 that uses only local variables 是定语从句修饰 function; rely on 意思是"依靠、依赖"; side effects 表示"副作用，不良反应"。

[3] Unlike most high-level computer languages, C imposes few restrictions on what you can do with it.

译文：不像其他高级计算机语言，C 语言对程序员能用它做什么没有什么限制。

说明：句中 what you can do with it 是宾语从句；impose on 意思是"把……强加于……；利用；欺骗"。例如：

Think of the restrictions you impose on yourself, and see if you can lift a few of them. 想一下你对自己的约束，看能不能解除一些。

The only limits to your success are those that you impose on yourself. 唯一限制你成功的因素都是你强加于自己的。

Exercises

Ⅰ. Answer the following questions according to the passage.
  1. What is the distinguishing feature of block-structured language?
  2. What is a block of code?
  3. What is one reason for the invention of C?

Ⅱ. Translate the following phrases into Chinese or English.
  1. 模块—结构化语言 _____
  2. 编程语言 _____
  3. 汇编语言 _____
  4. blocks of code _____
  5. curly brace _____
  6. side effects _____

Ⅲ. Fill in the blanks with the proper word. Change the form if necessary.

| mode | output | microcomputers | speed |
| input | digital | equal | data |

We can define a computer as a device that accepts _____, processes _____, stores data, and produces _____. According to the _____ of processing, computers are either analog or _____.

They can also be classified as mainframes, minicomputers, workstations, or _____ . All else (for example, the age of the machine) being _____, this categorization provides some indication of the computer's _____, size, cost, and abilities.

IV. Translate the following paragraph into Chinese.

A programming language consists of all the symbols, characters, and usage rules that permit people to communicate with computers. Some programming languages are created to serve a special purpose machine, while others are more flexible general-purpose machines that are suitable for many types of applications.

Translating Skills

长难句的翻译

所谓长句是指结构复杂，往往有多层并列或主从关系的句子；所谓难句是指看上去似乎易懂，但用中文表述起来却很难的句子。长难句在英语科技文献或电子产品应用手册中出现的频率很高，它也是阅读中较难的一部分。掌握好长难句的翻译方法和技巧对电子信息技术的获取和学习尤为重要。一般在英文科技资料里面可能会有多个从句，从句与从句之间的关系可能为并列、包含与被包含、镶嵌等形式。因此分析长难句或者翻译长难句，首先应该弄清楚句子的主干，从句的结构以及各从句之间的关系。经常出现的复合语句包括名词性从句（它又包括主语从句、宾语从句、表语从句和同位语从句）、形容词性从句（即定语从句），及状语从句。

分析长难句有以下几点最为基础的步骤：
1) 找出全句的主语、谓语和宾语，即句子的主干结构。
2) 找出句中所有的谓语结构、非谓语结构、介词短语和从句的引导词。
3) 分析从句和短语的功能，例如，是否为主语从句、宾语从句、表语从句或状语从句等，分析词、短语和从句之间的关系。
4) 分析句子中是否有固定词组或固定搭配、是否有插入语等其他成分。

通常解析长难句的方法有以下5种：顺序法、逆序法、包孕法、分句法、综合法。

（1）顺序法

当英语长句的内容叙述层次与汉语基本一致时，可以按照英语原文表达的层次顺序翻译，从而使译文与英语原文的顺序基本一致。例如：

But now it is realized that supplies of some of them are limited, and it is even possible to give a reasonable estimate of their "expectation of life", the time it will take to exhaust all known sources and reserves of these materials.

分析：该句的骨干结构为"It is realized that ...", it 为形式主语，that 引导主语从句以及并列的 it is even possible to... 结构。其中，不定式作主语，the time... 是"expectation of life"的同位语，进一步解释其含义，而 time 后面的句子是它的定语从句。5个谓语结构表达了4个层次的意义：①可是现在人们意识到；②其中有些矿物质的蕴藏量是有限的；③人们甚至还可以比较合理地估计出这些矿物质"可望存在多少年"；④这些已知矿源和储量将消耗殆尽的时间。根据同位语从句的翻译方法，把第四层意义的表达作适当的调整。

译文：可是现在人们意识到，其中有些矿物质的蕴藏量是有限的，人们甚至还可以比较合理的估计出这些矿物质"可望存在多少年"，也就是说，经过若干年后，这些矿物的全部已知矿源和储量将消耗殆尽。

(2) 逆序法

英语有些长句的表达次序与汉语表达习惯不同，甚至完全相反，这时必须从原文后面开始翻译。在汉语中，定语修饰语和状语修饰语往往位于被修饰语之前；在英语中，许多修饰语常常位于被修饰语之后，因此翻译时往往要把原文的语序颠倒过来。倒置法通常用于英译汉，即对英语长句按照汉语的习惯表达法进行前后调换，按表达含义全部倒置，原则是使汉语译句符合现代汉语论理叙事的一般逻辑顺序。例如：

The construction of such a satellite is now believed to be quite realizable, its realization being supported with all the achievements of contemporary science, which have brought into being not only materials capable of withstanding severe stresses involved and high temperatures developed, but new technological processes as well.

分析：该句由主句、作原因状语的分词独立结构、修饰独立结构的定语从句3部分构成。根据汉语词序，状语，特别是原因状语在先，定语前置，故从which... 入手，再译出 its realization...，最后才译出 The construction... realizable。

译文：现代科学的一切成就不仅提供了能够承受高温高压的材料，而且也提供了新的工艺过程。依靠现代科学的这些成就，我们相信完全可以制造出这样的人造卫星。

(3) 包孕法

这种方法多用于英译汉。所谓包孕是指在把英语长句译成汉语时，把英语后置成分按照汉语的正常语序放在中心词之前，使修饰成分在汉语句中形成前置包孕。但修饰成分不宜过长，否则会形成拖沓或造成汉语句子成分在连接上的纠葛。例如：

You are the representative of a country and of a continent to which China feels particularly close.

译文：您是一位来自于使中国倍感亲切的国家和大洲的代表。

What brings us together is that we have common interests and technology which transcend those differences.

译文：使我们走到一起的，是我们有超越这些分歧的共同利益和技术。

(4) 分句法

有时英语长句中主语或主句与修饰词的关系并不十分密切，翻译时按照汉语多用短句的习惯，把长句的从句或短语转换成句子，分开来译。为使语意连贯，有时需适当增加词语，也就是采取化整为零的方法将整个英语长句翻译为几个独立的句子，顺序基本不变，前后保持连贯。如：

Television, it is often said, keeps one informed about current events, allow one to follow the latest developments in science and politics, and offers an endless series of programs which are both instructive and entertaining.

分析：在此长句中，有一个插入语"it is often said"，3个并列的谓语结构，还有一个定语从句，其中3个并列的谓语结构尽管在结构上属于同一个句子，但都有独立的意义。因此在翻译时，可以采用分句法，按照汉语的习惯把整个句子分解成几个独立的分句。

译文：人们常说，通过电视可以了解时事、掌握科学和政治的最新动态。从电视里还可以看到层出不穷、既有教育意义又有娱乐性的新节目。

(5) 综合法

上面我们讲述了英语长句的逆序法、顺序法、包孕法和分句法。事实上，在翻译一个英语长句时，并不单纯地使用一种翻译方法，而是要综合使用各种方法，这在我们上面所举的例子中也有所体现。再如，一些英语长句单纯采用上述任何一种方法都不方便，这就需要我们仔细分析，或按照时间的先后，或按照逻辑顺序，顺逆结合，主次分明地对全句进行综合处理，以便把英语原文翻译成通顺忠实的汉语句子。例如：

From the uniformity of slight pressure, we can be aware how deeply a finger is thrust into water at body temperature, even if the finger is enclosed in a rubber glove that keeps the skin completely dry.

分析：该句共有以下3层含义：①根据均匀的微小压力我们可以知道手指伸进水中的深度；②水温与体温相同；③手指被能使皮肤保持干爽的橡胶手套包围。在这3层含义中，③表示条件和让步，②表示条件，而①则表示结果。

译文：当水温与体温相同时，根据均匀的微小压力，即使戴着能使皮肤保持干爽的橡胶手套，我们也知道手指伸进水中的深度。

Reading

MATLAB Language

MATLAB is a numerical computing environment and programming language. The MATLAB integrates computation, visualization, and programming in an easy-to-use environment where problems and solutions are expressed in familiar mathematical notation.[1]

MATLAB is an interactive system whose basic data element is an array. It allows you to solve many technical computing problems, especially those with matrix and vector formulations. An additional package, Simulink, adds graphical multidomain simulation and Model-Based Design for dynamic and embedded systems.

MATLAB is easy to use, here are two examples.

(1) Start & quit MATLAB

When you start MATLAB, the desktop appears containing tools (graphical user interfaces) for managing files, variables, and applications associated with MATLAB.

Fig. 15-1 shows the default desktop. You can customize the arrangement of tools and documents to suit your needs.

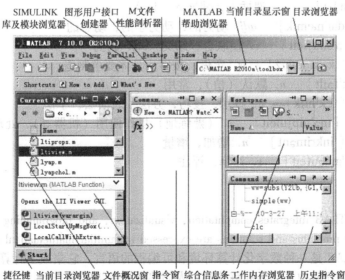

Fig. 15-1 Desktop of MATLAB

To end your MATLAB session, select File → Exit MATLAB in the desktop, or type quit in the Command Window.

(2) Plotting Process

The MATLAB environment provides a wide variety of techniques to display data graphically. Interactive tools enable you to manipulate graphs to achieve results that reveal the most information about your data.

For example, the following statement creates a variable x that contains values ranging from -1 to 1 in increments of 0.1. The second statement raises each value in X to the third power and stores these values in y:

```
x = -1..1;              % define x ray
y = x.^3;               % define the third power of each value in x and store as y
plot (x, y);            % draw a curve of y-x
```

A simple line graph (seen in Fig. 15-2) is a suitable way to display x as the independent variable and y as the dependent variable.

You can also annotate and print graphs for presentations, or export graphs to standard graphics formats for presentation in Web browsers or other media.

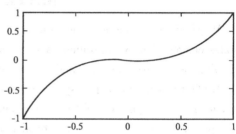

Fig. 15-2 A Simple Line Graph

New Words and Phrases

numerical   [njuː'merikəl]   adj. 数字的，用数表示的
computation [ˌkɔmpjuː'teiʃən] n. 计算，估计
interactive [ˌintər'æktiv]   adj. 交互式的；相互作用的
array   [ə'rei]   n. 数组，阵列；排列 vt. 排列
formulation [ˌfɔːmjuː'leiʃən] n. 构想，规划；公式化
embed   [im'bed]   vt. 使插入，使嵌入，嵌入
dynamic   [dai'næmik]   adj. 动力的，动力学的，动态的
associate   [ə'səuʃieit]   v. 使联合，结交 adj. 副的
customize   ['kʌstəmaiz]   vt. 定做，按客户具体要求制造
default   [di'fɔːlt]   n. 默认（值），常用
technique   [tek'niːk]   n. 技术，技巧，方法，表演法，手法
manipulate   [mə'nipjuleit]   v.（熟练地）操作使用（机器等），处理
increment   ['inkrimənt]   n. 增加，增量
annotate   ['ænəuteit]   v. 注释，评注

Notes

[1] The MATLAB integrates computation, visualization, and programming in an easy-to-use environment where problems and solutions are expressed in familiar mathematical notation.

译文：MATLAB 是一种数字化计算环境和可编程的语言。它集成了计算、可视化和编程在一个极易使用的环境中，问题和解答都以人们熟悉的数学符号表示。

说明：本句是复合句。句中 where problems... notation 是定语从句，修饰 environment。

Translate the following sentences into Chinese.

1. This choice was made because it provides easy-to-use editors and tooling for analysis engine development.
2. To run it in interactive mode, type each command at the prompt.
3. These applications contain both the interactive aspects as well as the business logic.
4. You can, of course, customize the page as you like.
5. You can manipulate your background jobs from the command line using these labels.

Unit 16　Multimedia Technology

Text

1. What is Multimedia

　　Some elements of media are animation, sound, graphics, text, video and photography.

　　Multimedia involves the combination of two or more media types to effectively create a sequence of events that will communicate an idea, usually with both sound and visual support. Typically, multimedia productions are developed and controlled by computer.

　　Multimedia isn't new, and the term has been used for decades to describe slide presentation accompanied by audio tape (slide/tape). The combination of slide and narration has been both a popular and successful form of business presentation.

　　In the 1970s the slide show format was introduced to the computer, this technology allowed the computer to control numerous projectors, coordinating them in a manner that produced fast-paced dissolves and effects. [1] Taped soundtracks would contain cues that triggered the slide projectors to do what it was programmed to do.

　　In the 1980s PCs were designed to "cut" a graphic element and "paste" it into another document. Since then software and hardware developers have been scrambling to integrate various forms of media into the personal computer.

2. Multimedia Assets

　　Effective multimedia applications depend on the most effective use of various materials referred to as assets or resources. There are various multimedia resources.

　　(1) Text

　　There are 3 major advantages generally associated with screen-based text compared with paper-based text: the ability to spontaneously update the screen, the reactive capability, and the ability to incorporate special effects.

　　(2) Audio

　　Interactive audio can add a particular dimension of reality to multimedia systems. Until recently, audio-based subject domains—such as music, linguistics, languages etc. have been virtually ignored. Also, the efficiency of interaction between the human and the machine could have been greater had sonic interaction and audio enhancements been available sooner. [2]

　　(3) Pictures

　　When using images the designer must decide on the most economical and efficient method to generate them. At the planning stage of the application the designer must decide the most appropriate combination of technology and software to be able to reproduce and create new images.

　　(4) Video Images

　　Portable video recorders have made it relatively easy to capture real-time video images. Video can be incorporated into multimedia applications using two different processes. The first involves using a video source connected to the computer via a controller card. This technology has been referred to as interactive video. A more integrated process converts video from analogue into digital format that can be manipulated by the desktop computer.

3. Multimedia Applications

There are various ways to group multimedia titles. They can be classified by market (such as home, business, government and school), by user (such as child, adult, teacher and student), or by category (such as education, entertainment and reference). Here we group them into seven application areas and to look at each in a more detail.

(1) Reference

Encyclopedias, census data, yellow pages, atlases and street directories are examples of CD reference titles. In many cases they are electronic versions of reference books. The challenge for the developer is to make it easy for the user to find the desired information and to effectively use other multimedia elements such as sound, video and animation.

(2) Education

The goal of the educator is to facilitate learning—to help the student gain a body of knowledge, acquire specific skills and function successfully in society. But one of the greatest challenges to educators is the diversity of students, especially in the different ways they learn. Some students learn better through association, others by experimentation, some are more visually oriented, others are more auditory.

Multimedia has the ability to accommodate different learning styles and can present material in a non-linear manner. It is motivating, it can be highly interactive, and it can provide feedback and evaluate skills.

(3) Training

Every company has a need to train its employees on a wide range of subjects from personnel policy to equipment maintenance.

With multimedia the trainee can perform a simulated job function in order to develop an advanced level without having touched the actual unit. The integration of audio and video allows this training technology to be a highly effective medium in areas such as flight and driving simulators. Similarly, NASA uses multimedia extensively for flight control training for astronauts.

(4) Entertainment

Drawing the line between education and entertainment in multimedia can be almost impossible, hence the term "edutainment". Multimedia can make learning entertaining.

But multimedia also has a purely entertainment side. Anything that's possible in sound and images is possible on a multimedia CD. AIATSIS is an encyclopedia of the Australian Aborigine containing over 2000 entries—1000 photos, 230 sound clips and 50 videos. It covers subjects ranging from art to health, from technology to law.

(5) Businesses

As businesses have the need to communicate with the outside world, multimedia processes offer a wide variety of options for business presentations, marketing and sales. Multimedia can be used at trade shows or to produce electronic catalogues. The marketing of new products can be greatly enhanced by using multimedia, these products can be marketed in a manner that will provide more detailed and stimulating information than printed media.

(6) Presentation

Thousands of multimedia presentations are made in the business world every day. Company

CEOs give their annual report to a meeting of stockholders. Sales representatives pitch their product line to a group of potential customers. A conference speaker tells an audience about industry trends. From an electronic slide show to an interactive video display multimedia can enhance a presentation.

Multimedia provides the presenter with the tools to attract and focus the audience's attention, reinforce key concepts and enliven the presentation.

(7) Interactive Game

Multimedia means interaction, and to many interactive entertainment means games. Game developers were the pioneers in the use of multimedia and still provide the most innovative and interactive applications of multimedia.

In order to attract, engage, captivate and challenge the user multimedia provides the fast action, vivid colors, 3D animations and elaborate sound effects that are essential to entertainment. It can also provide the rewards, recognition and sense of accomplishment that are often part of entertainment titles.

Many games have moved from the physical (hand/eye coordination) to the mental (solving the mystery, overcoming evil, outwitting the opponent).

On the other hand, hobbies and sports are examples of multimedia titles that provide the user with a vicarious experience such as being able to play the best golf courses in the world or simulate flying over 3D cityscapes.

New Words and Phrases

assets ['æsets] n. 资产；有用的东西；有利条件；优点
aborigine [æbə'ridʒiniː] n. 土著，土著居民
Aborigine n. 澳大利亚土著居民
cityscape ['sitiskeip] n. 都市风景
coordinate [kəu'ɔːdinit] v. 协调，综合 n. 同位格 adj. 同等的
scramble ['skræmbl] vi. 攀登；仓促行动 vt. 攀登 n. 爬行，攀登
spontaneous [spɔn'teiniəs] adj. 自然的，不由自主的 adv. 自发地，天真地
sonic ['sɔnik] adj. 音速的；声音的；声波的
multimedia [mʌlti'miːdiə] n. 图文、视频和音频等的合成 adj. 多媒体的
animation [æni'meiʃn] n. 动画，活跃；生气勃勃；兴奋
dissolve [di'zɔlv] n. 渐渐消隐，溶化 vt. 叠化，叠化画面
encyclopedia [inˌsaiklə'piːdiə] n. 百科全书
encyclopedias [ˌensaiklou'piːdjəs] adj. 如百科辞典的，百科全书式的
atlas ['ætləs] n. 地图集
incorporate [in'kɔːpəreit] adj. 合并的，一体化的 vt. & vi. 合并，具体表现
manipulate [mə'nipjuleit] vt. （熟练地）操作，使用（机器等），巧妙地处理
outwit [ˌaut'wit] vt. 瞒骗，以智取胜
projector [prə'dʒektə (r)] n. 电影放映机；幻灯机；投影
soundtrack ['saundtræk] n. 声带；声迹；音轨；声道；电影配音
trigger ['trigə (r)] vt. 引发，引起，触发 n. 扳机

Notes

[1] In the 1970s the slide show format was introduced to the computer, this technology allowed the computer to control numerous projectors, coordinating them in a manner that produced fast-paced dissolves and effects.

译文：在20世纪70年代，幻灯播放方式被引入到计算机中，允许计算机控制多个投影仪，协调它们产生一种快节奏的切换方式和生动的效果。

说明：句中coordinating作伴随状语；fast-paced dissolves表示"快速切换"。

[2] Also, the efficiency of interaction between the human and the machine could have been greater had sonic interaction and audio enhancements been available sooner.

译文：如果能更快地实现声音的交互和音频（质量）的改进，则人和计算机之间的交流将变得更加生动。

说明：本句使用了虚拟语气，had sonic interaction and audio enhancements been available sooner是虚拟条件句。

Exercises

I. Translate the following phrases into English.

1. 音频
2. 多媒体
3. 找到想要的信息
4. 视频
5. 幻灯片
6. 提供反馈（信息）
7. 娱乐
8. 三维动画

II. Translate the following phrases into Chinese.

1. create a sequence of events
2. multimedia assets
3. portable video recorders
4. digital video
5. learn better through association
6. has a purely entertainment side
7. the most economical and efficient method
8. the most appropriate combination of technology and software

III. Translate the following passage into Chinese.

Four Stages of Circuit Simulation

The simulator in Electronics Workbench (EWB), like other general-purpose simulators, has four main stages: input, setup, analysis and output.

At the input stage, after you have built a schematic, assigned values and chosen an analysis, the simulator reads information about your circuit.

At the setup stage. the simulator constructs and checks a set of data structures that contain a complete description of your circuit.

At the analysis stage, the circuit analysis specified in the input stage is performed. This stage occupies most of CPU execution time and actually is the core of circuit simulation. The analysis stage formulates and solves circuit equations for the specified analyses and provides all the data for direct output or post-processing.

At the output stage, you view the simulation results. You can view results on instruments such as the oscilloscope, or on graphs that appear when you run an analysis from the Analysis menu or when you chooseAnalysis/Display Graphs.

Translating Skills

电子产品的英文说明书

电子产品说明书用来对电子产品进行介绍和说明,包括产品的外观、性能、参数、使用方法、操作指南和注意事项等,具有传播知识、指导消费、宣传企业等作用,所以书写电子产品说明书必须采用客观、准确的语言。此外,也可以采用图文兼备等多种形式来描述,以便指导消费者使用和维修产品。依据电子产品的大小、复杂程度等不同形式,产品说明书的内容又略有差异。同样,电子产品说明书英译时必须要准确(accuracy)、简明(conciseness)、客观(objectivity)。

(1) 准确

专业术语、固定用语和习惯说法必须表达得准确、恰当。例如,在翻译数码相机说明书时会遇到这样一些术语:镜头后盖(ear lens cap)、三角架(tripod)、数码变焦(digital zoom)、快门帘幕(shutter curtain)、曝光不足(under exposure)、取景器(view finder)等,需按专业说法表达出来,不可臆造。

(2) 简明

1)内容条目要简洁明了、步骤清晰、逻辑性强。例如,部件名称、操作界面等都可以配以示意图,再用箭头注明;操作步骤等可以用项目符号或编号依次标出;有些地方还把数据信息列成表格,使人一目了然。

2)可以使用常用的缩略形式,如液晶显示器(Liquid Crystal Display)常缩写成 LCD、发光二极管(Light Emitting Diode)常缩写成 LED、中央处理器(Central Processing Unit)常缩写成 CPU。

(3) 客观

电子产品说明书必须将该产品的相关内容客观地呈现出来。

为了使说明书表达准确、简明易懂又不失客观,可通过如下方式来实现。

1)广泛使用复合名词结构。

在译文中可以用复合名词结构代替各种后置定语,以求行文简洁、明了、客观。例如,原文中使用"设备清单",译文可以用 equipment check list(不用 the list of equipment check);原文中使用"保修卡",译文可以用 warranty card(不用 the card of warranty)。

有时候一些小标题常英译成动名词短语。例如,"测光模式"可以译为 metering modes。译句可以使用非人称名词化结构作主语,使句意更客观、简洁。例如,原文为"由于使用了计算机,数据计算方面的问题得到了解决",可以译为"The use of computers has solved the problems in the area of calculating"。

2)普遍使用一般现在时。

电子产品说明书的主体部分一般采用"无时间性"(Timeless)叙述,其译文普遍使用一般现在时,以体现内容的客观性。例如,原文为"本传真机与数码电话系统不兼容。"可以译为"This facsimile machine is not compatible with digital telephone systems."

3)常使用被动语态。

被动语态可以使译文客观简洁,而且可以使读者的注意力集中在受动者这一主要信息上。例如,原文为"您可以在光盘中的电子使用手册中找到额外的信息。"可以译为:Additional information can be found in the electronic user's manual which is located on the CD-ROM.

4)广泛使用祈使句。

祈使句使译文的表述显得准确、客观又简洁。例如，Turn off your computer and unplug its power cable. 确定关闭计算机主机电源并将电源插头拔出。

以下是 Super Pointer R SP-100 电子产品说明书中的一部分。

Super Pointer R SP-100

1. **Product composition**（seen in Fig. 16-1）
2. **General Description & Key Features**

   (1) General Description
   - To make your presentations flow continuously, the Super Pointer integrates the slide up/down functions with a laser pointer on a remote hand held device using a wireless technology. The Super Pointer's attractive and compact design enables mobility of up to Meters from the laptop/PC while still maintaining a full navigational control of your presentations.
   - The Super Pointer does not require any software and PC programming and take just a few seconds to install and use.

   (2) Key Features
   - Plug and play function (No PC programming).
   - Working on any computer (On any computer using a USB port).
   - Remote control range over 15m with a very low power (Eliminates LOS problems encountered with IR systems).
   - Transmitter identification (Use thousands of ID number to avoid multi-user interferences)

   (3) Terms and Explanations

   1) LED indicator. It blinks when connected and waiting for ID. It is off after ID is received and waiting for command. It is off-on-off when every command is received.

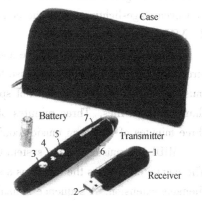

Fig. 16-1   Product Composition

   2) USB connector.
   3) Laser button: laser on/off control.
   4) Up button: slide down control.
   5) Down button: slide up control.
   6) Lock/Unlock switch: to change battery.
   7) Battery cover.

3. **Installation and Usage**

   1) Plug the receiver into USB port. Then the LED indicator will start to blink.
   2) Push either the down or up button for 1-2 seconds. The LED indicator will turn off.
   3) Begin you Power Point presentation.

4. **Technical Specifications & Installation and Usage**
   - Technical Specifications (omitting)
   - Installation and Usage: 1) Lock/Unlock Switch; 2) Push Battery Cover.

5. **Special Notice**
   - Do not aim the super pointer to the eye directly. It may damage the eye.
   - The super pointer can not be sold to children under 19 years old.

6. **Remark**
   - In case of malfunction, pull out and install again using a receiver or reboot a computer.

WARRANTY

This product is warranted against defect for one year from the date of purchase. Within this period, simply take the product and your proof of purchase any XX dealer and the product will be repaired or exchanged without charge for parts and labor. Any product which has been subject to misuse or accidental damage is excluded from this warranty.

Reading

MIDI Interface

1. MIDI

MIDI is short for Musical Instrument Digital Interface, a standard specification developed by music synthesizer manufacturers. The concept of being able to control several instruments from one keyboard has grown into a method for putting musical instruments, tape recorders, VCRs, mixers, and even stage-lighting under the control of a single computer.

2. MIDI Music

Can I use MIDI to store voice and music data? MIDI files contain the instructions that MIDI instruments and MIDI sound cards use to recreate or synthesize sounds. MIDI files store and recreate musical instrument sounds, but not speaking or singing voices. MIDI files are much more compact than waveform files. Three minutes of MIDI music requires only 10KB of storage space, whereas three minutes of waveform music requires 15MB.

MIDI is a music notation system that allows computers to communicate with music synthesizers. The computer encodes the music as a sequence and stores it as a file with a . mid, . cmf, or . rol filename extension. A sequence is analogous to a player-piano roll that contains punched information indicating which musical notes to play. A MIDI sequence contains instructions specifying the pitch of a note, the point at which a note begins, the instrument that plays the note, the volume of the note, and the duration of the note.

Most of today's computers include a MIDI-capable sound card. You can download MIDI music files from the Internet and play them on your computer. You can also plug a MIDI instrument, such as an electronic keyboard, into the sound card's MIDI/joystick port to record your own MIDI tunes.

A MIDI sequence is stored as a series of tracks. Each track represents an instrument. When you compose MIDI music, you can assign an instrument to each track, write the notes the instrument should play, and indicate the volume and sound quality.

New Words and Phrases

    interface   [ˈintəfeis]   n. 接口；界面；接触面
    synthesizer   [ˈsinθisaizə]   n. 音响合成器，综合器，合成器
    megabytes   [ˌmegəˈbaits]   n. 兆字节，可简写为 MB
    encode   [inˈkəud]   n. 编码   vt. 把……编码
    punch   [pʌntʃ]   v. 打孔   n. 冲压机；钻孔器   vt. 开洞；以拳重击   vi. 用拳猛击
    pitch   [pitʃ]   vt. 定调，定位于；掷   vi. 投掷；倾斜   n. 程度；音高；投掷
    joystick   [ˈdʒɔistik]   n. 游戏操纵杆，控制杆

Exercises

I. **Answer the following questions according to the reading.**

1. What is an MIDI Interface?
2. What should we do when we play or record a music?
3. How can we compose MIDI music?

II. **Translate the following sentences into English.**

1. MIDI 是由音乐合成器制造商开发的标准规范。
2. MIDI 文件比波形文件更紧凑。
3. 计算机把音乐编码成一个序列，并以 .mid、.cmf 或者 .rol 的文件扩展名进行存储。

Unit 17 China's Progress in Supercomputing

Text

China's Tianhe-1A supercomputer, developed by China's National University of Defense Technology, has taken top spot among the world's top 500 supercomputers. This is an important marker for the rapid growth of China's science and technology.

The Tianhe-1A system at the National Supercomputer Center in Tianjin can perform a mind-numbing 2.57 quadrillion calculations per second.

Among the top 500 computers on the list, officially unveiled on Tuesday by researchers from the United States and Europe, 41 are from China. China also has two systems in the top 10, of which seven achieved performance at or above one petaflop/s. The US has 275 computers on the list; France, Germany and Japan each have 26, Britain 24 and Russia 11.

It is a significant achievement for China, given the fierce international competition in supercomputing, which is a fundamental tool for a nation's scientific research in various areas, such as geology, meteorology, and oil exploration as well as the aviation, automobile and chemical industries. [1]

The list was compiled by Hans Meuer of the University of Mannheim, Germany, Erich Strohmaier and Horst Simon of the National Energy Research Scientific Computing Center (NERSC)/Lawrence Berkeley National Laboratory, and Jack Dongarra of the University of Tennessee, Knoxville.

In the early 1990s, almost all computers in China were imported and scientific research institutes could only use computers under the supervision and control of foreign experts. In recent years, Chinese researchers have worked hard to develop China's own supercomputers, with the support of the 863 Program, a government-sponsored national high-tech research and development initiative.

These "Made-in-China" supercomputers not only break the blockade imposed by foreign countries, they also narrow the gap between China and developed countries. [2] They have even made China a world leader in some technologies.

To some extent, supercomputers reflect a country's demand for computing capacity and the development level of a country's fundamental research and development and high-tech industries. As an official from the French Atomic Energy Commission pointed out the computing speed of Tianhe-1A showed the increased competitive strength of China's economy. However, it needs to be remembered that China still lags behind such developed countries as the US in overall computing capacity, especially in core electronics and high-end chip development. [3]

The US and Japan are developing supercomputers with better performance than the Tianhe-1A, and in the future, various countries will probably take turns holding the lead in supercomputer development. China still faces an arduous task catching up with supercomputing in other countries.

However, Tianhe-1A has given "Made-in-China" its place in the realm of supercomputers and sent a signal announcing the rapid development of China's science and technology.

New Words and Phrases

quadrillion [kwɔːˈdriljən] n. 千的五次方，百万的四次方；万亿

unveil	[ʌn'veil]	vt. 使公之于众，揭开；揭 vi. 除去面纱；显露
petaflop		千万亿次
performance	[pə'fɔ:m(ə)ns]	n. 性能；绩效；表演；执行；表现
achievement	[ə'tʃivmənt]	n. 完成，达到；成就，成绩
geology	[dʒi'ɔ:lədʒi]	n. 地质学；（某地区的）地质情况
meteorology	['mi:tiə'rɔ:lədʒi]	n. 气象学
exploration	[eksplə'reiʃ(ə)n]	n. 探测；探究；踏勘
compile	[kəm'pail]	vt. 编译；编制；编辑
sponsor	['spɔnsə]	n. 赞助者；主办者；保证人 vt. 赞助；发起
blockade	[blɔ:'keid]	n. 封锁；封锁部队；障碍物，阻碍物
narrow	['nærəu]	adj. 狭窄的；勉强的 n. 海峡；隘路 vt. 使变狭窄 vi. 变窄
impose	[im'pəuz]	vi. 利用；欺骗；施加影响 vt. 强加；征税；以……欺骗
initiative	[i'niʃətiv]	n. 主动权；首创精神的主动行动；提倡 adj. 主动的；自发的
competitive	[kəm'petitiv]	adj. 竞争的，比赛的；（价格等）有竞争力的
capacity	[kə'pæsiti]	n. 能力；容量；资格，地位；生产力
chip	[tʃip]	vt. 削，凿；削成碎片 vi. 剥落；碎裂 n. 电子芯片；碎片
lag	[læg]	n. 落后；迟延 vt. 落后于 vi. 滞后；缓缓而行；蹒跚 adj. 最后的
arduous	['ɑ:djuəs]	adj. 努力的；费力的；险峻的
realm	[relm]	n. 领域，范围；王国
mind-numbing	['maindnʌmbiŋ]	adj. 令人心烦意乱的；令人厌恶的；无法想象的
high-end	['haiend]	adj. 高端的；高档的
catch up with		赶上，追上；逮捕；处罚

Notes

[1] It is a significant achievement for China, given the fierce international competition in supercomputing, which is a fundamental tool for a nation's scientific research in various areas, such as geology, meteorology, and oil exploration as well as the aviation, automobile and chemical industries.

译文：考虑到超级计算机领域的激烈竞争，这是中国的一次巨大成功。超级计算机是一个国家在很多领域进行科学研究的基础工具，这些领域包括地理、气象和石油勘探以及航空、汽车、化工等。

说明：这是一个复合句。主句 It is a significant achievement for China 是主系表结构，it 指代中国天河1号；句中 given 是"考虑到"的意思，在这里引出条件状语；which 引导的非限定性定语从句修饰 supercomputing。

[2] These "Made-in-China" supercomputers not only break the blockade imposed by foreign countries, they also narrow the gap between China and developed countries.

译文：这些"中国制造"的超级计算机不仅打破了外国的封锁，也缩小了中国与发达国家的差距。

说明：这是一个并列句。英语中 not only... but also 主要用于连接两个对等的成分，句子中省略了 but，但在 also 前面加了主语 they 指代前面的 supercomputers，这是一种灵活的写作手法；impose 是"施加，强加"的意思，根据汉语表达习惯，翻译时可省略。例如：

And then provides the force perpendicular to the direction along the back pad by loading imposed on the load point. 再将规定的力沿着与椅背垂直的方向通过加载垫施加于加载点上。

[3] It needs to be remembered that China still lags behind such developed countries as the US in overall computing capacity, especially in core electronics and high-end chip development.

译文：需要记住并引起注意的是，在整体计算能力方面，尤其是核心电子和高端芯片的发展方面，中国仍落后于像美国这样的发达国家。

说明：这是一个复合句。It 是形式主语，that 引导的主语从句太长而放在后面是为了避免头重脚轻；lag behind 表示"落后于……"的意思，如：

As catalysts of new ideas, German universities lag behind American and British ones. 另外，作为新想法的催化剂，德国大学也落后于英美大学。

Exercises

Ⅰ. Answer the following questions according to the test.

1. What it the calculation speed of the Tianhe-1A system at the National Supercomputer Center in Tianjin?
2. At what time that almost all the computers in China are imported?
3. In which fields thatChina's especially lags behind developed countries regarding supercomputer development?
4. Which countries are developing supercomputers with better performance than the"Tianhe-1A" super computer?

Ⅱ. Match the following phrases in column A with column B.

Column A	Column B
1. petaflop	a. 性能，表现
2. initiative	b. 能力，产能
3. capacity	c. 千万亿次
4. high-end	d. 追上
5. catch up with	e. 主办者；赞助商
6. performanc	f. 无法想象的
7. mind-numbing	g. 高端的
8. sponsor	h. 主动权；首创精神的主动行动

Ⅲ. Fill in the blanks with the proper word. Change the form if necessary.

| announce | that | rank | simulation |
| run | until | overtake | achieve |

Tianhe-IA was _____ as the world's fastest supercomputer in the TOP500 list _____ July 2011 when the K computer _____ it. In June 2011, scientists at the Institute of Process Engineering (IPE) at the Chinese Academy of Sciences (CAS) _____ a record-breaking scientific _____ on the Tianhe-1A supercomputer _____ furthers their research in solar energy. CAS-IPE scientists _____ a complex molecular dynamics simulation on all 7168 NVIDIA Tesla GPUs to _____ a performance of 1.87 petaflops (about the same performance as 130,000 laptops).

Ⅳ. Translate the following sentences into Chinese.

1. It consumes 4.04 megawatts of power in operation, covers 17,000 square feet comprising 103 computer racks.
2. News media and commentators talked about how the U.S. had been dethroned after six years

of having its systems top the top500 list.

3. It claims a performance record of 2.507 petaflops, smashing a previous record set by the Cray XT5 Jaguar.

Useful Information

专业网站介绍与英文网站注册申请表的填写

1. 电子信息专业的常用英文网站

作为一个合格的电子工程师，通过网站获得技术资料、找到自己所需要的产品和器件和学习提高是必须掌握的本领。下面分别介绍最著名的几个的电子器件贸易网站和电子信息英文网站。

（1）欧时电子 www.rs-online.com

欧时是世界著名的电子元器件、连接器、电容器、微处理器、开发工具包、电阻器、测试和测量仪器分销商，产品来自众多世界顶级制造商。开发工具包括很多原理图、大量器件的各种格式的 2D 和 3D 数字模型，而这些多数是免费的、方便工程师的设计工作。

（2）伍尔特电子 http://www.we-online.com

伍尔特是一家源自德国的综合工业公司，是全球装配和紧固件业务市场领导者，旗下有一个专门的电子专业门类销售部门。

（3）米思米 http://jp.misumi-ec.com/top/

米思米是源自日本的全球专业零配件供应商，涵盖了 FA 工厂自动化、冲压/塑料模具、电子电气、工具/MRO 工厂消耗品等各种高质量的零件，6 大目录免费申请。旗下有一专业电子器件销售部门。站内还提供大量的器件规格、2D 和 3D 数字模型甚至成熟的设计结构 3D 数据供下载。

（4）贸泽电子 http://www.mouser.com/

贸泽电子（Mouser Electronics），总部位于美国得克萨斯州的曼斯菲尔德，是半导体与电子元器件业的全球分销商，获得 500 多家知名厂商授权分销其产品。除了在该网站寻找并采购自己所需要的电子器件外，还可以登录其博客网页，参考大量专业人士的博客文章。

（5）电子工程专辑 http://www.eetimes.com/

电子工程专辑有非常新的国际资讯，它有中国国内专辑，网站的内容十分丰富。

（6）电子设计技术 www.edn.com

EDN 隶属于全球著名的媒体集团 Reed 集团，拥有专业的媒体资源，在国内拥有较高的知名度。EDNChina 是中国国内合作方，资讯更新比较快，而且技术方案也比较齐全，它的一个特色是博客做得不错，有很多精彩的内容。

（7）电子系统设计 http://www.edn.com/

电子系统设计网站是美国著名的专业媒体，其内容风格是突出实用设计，网站有很多实用性很强的技术方案和设计技巧，是初学者学习提高的好地方。另外，这个网站的论坛内容也很丰富，对刚入门的工程师来说很有帮助。

2. 英文网站的注册方法

有时为了获取更多的资料，用户常需要在一些网站注册成为会员，因此就涉及注册申请表的填写。英文网站注册申请表的填写过程中涉及许多内容，下面以 eBay 的注册为例列出主要的步骤。

Log on the website of eBay (www.eBay.com) and press the button "Register".

(1) Enter your personal data

Tell us about yourself (All fields are required)

First name: Last name:

Street address: City: State / Province:

ZIP / Postal code: Country or region:

Primary telephone number: ext. :   Example: 123-456-7890

(Telephone is required in case there are questions about your account.)

Email address:

Re-enter email address:

(We're not big on spam. You can always change your email preferences after registration.)

(2) Create your user ID and password

Choose your user ID and password (All fields are required)

Create your eBay user ID:

1) In order to check your user ID can be used, please click "Check availability" button. If it is not available, please try again.

2) Use letters or numbers, but not symbols.

Create your password: (case sensitive)

Re-enter your password:

Pick a secret question: (Select your secret question...)

Your secret answer:

If you forget your password, we'll verify your identity with your secret question.

Date of birth: Month   Day   Year

(You must be at least 18 years old to use eBay.)

Foradded security, please enter the verification code hidden in the image.

Enter the verification code:

I agree that:

- I accept the User Agreement and Privacy Policy.
- I'm at least 18 years old.
- I may receive communications from eBay and can change my notification preferences in My eBay.

(3) Additional information from you

We need some additional information from you. Please check and try again.

Confirm your identity

Registration with eBay is free, but we need to verify your identity with a credit or debit card. This helps keep eBay a safe place to buy and sell.

(Don't worry, we won't charge your card and we won't share this information.)

Credit or debit card number: (Visa, MasterCard, American Express or Discover)

Expiration date: month: year:

Card identification number:

Your privacy is important to us.

eBay does not rent or sell your personal information to third parties without your consent. Your address will be used for shipping your purchase or receiving payment from buyers.

以下为注册信息的汉语译文

登录 eBay 的网站（www. eBay. com），单击"注册"按钮。

（1）输入个人数据

告诉我们一些你的情况（所有项必须填）

名字：　　　　　姓氏：

街道地址：　　　城市：　　　　　州/省：

邮政编码：　　　国家或地区：

主要的联系电话号码：　　　　分机号码：如 123 – 456 – 7890

（电话号码是必需的，一旦账户出现问题可以方便联系）

电子邮件地址：

再次输入电子邮件地址：

（我们并不热衷于发垃圾邮件，可以在注册后更改电子邮件参数设置。）

（2）创建用户名及密码

选择用户名和密码（所有项必须填写）

创建用户名：

1）为了测试用户名是否已可以使用，请单击"测试可用性"按钮。如果不可用，请更换用户名。

2）可以使用字母或数字，但不能用符号。

建立密码：（区分大小写）

再次输入密码：

选择一个秘密问题：（选择你的密码问题）

秘密问题答案：

（如果忘记了密码，我们会用秘密问题核实你的身份）

出生日期：　　　月　　　　年

（在 eBay 注册的必须是年满 18 周岁的成年人）

为增加安全，请输入隐藏在图片中的验证码。

输入验证码：

我同意下面的内容：

- 我接受用户协议和隐私政策。
- 我 18 周岁以上。
- 我接受来自 eBay 的通知，并可以改变我在 eBay 的通知参数设置。

（3）一些额外的信息

我们需要你一些额外的信息。请检查且再试一次。

确认你的身份。

注册 eBay 是免费的，但是我们需要凭借信用卡或借记卡来核实你的身份。这有助于维持易趣购买和出售的安全。

（别担心，我们不向你的信用卡收取费用，也不会共享这些信息。）

信用卡或借记卡号码：（Visa, MasterCard, American Express or Discover）

截止日期：　　　月　　　　年

卡识别号码：

你的隐私的对我们很重要。

eBay 不会未经你的同意将你的个人信息出租或出售给第三方。你的地址将被用来运送你所购物品或收取来自买家的付款。

Reading for Celebrity Biography (Ⅱ)

Bill Gates

William (Bill) H. Gates is chairman and chief software architect of Microsoft Corporation, the worldwide leader in software, services and Internet technologies for personal and business computing. [1] Microsoft had an income of $25.3 billion for the financial year ending June 2001, and employs more than 40,000 people in 60 countries.

Born on October 28, 1955, Gates and his two sisters grew up in Seattle. Gates attended public elementary school and the private Lakeside school. There, he discovered his interest in software and began programming computers at age of 13.

In 1973, Gates entered Harvard University as a freshman, where he lived down the hall from Steve Ballmer, now Microsoft's chief executive officer. While at Harvard, Gates developed a version of programming language BASIC for the first microcomputer—the MITS Altair.

In his junior year, Gates left Harvard to devote his energies to Microsoft, a company he had began in 1975 with his childhood friend Paul Allen. Guided by a belief that the computer would be a valuable tool on every office desktop and in every home, they began developing software for personal computers. Gates' anticipation and his vision for personal computing have been central to the success of Microsoft and the software industry. [2]

In 1999, Gates write Business at the Speed of Thought, a book that shows how computer technology can solve business problems in fundamentally new ways. The book was published in 25 languages and is available in more than 60 countries. Business at the Speed of Thought has received wide critical applause, and was listed on the best-seller lists of the New York Times, USA Today, the Wall Street Journal and Amazon. com. Gates' previous book, The Road Ahead, published in 1995, held the No. 1 place on the New York Times' bestseller list for seven weeks.

In addition to his love of computers and software, Gates is interested in biotechnology. He is an investor in a number of other biotechnology companies. Gates also founded Corbis, which is developing one of the world's largest resources of visual information. In addition, Gates has invested with cellular telephone pioneer Craig McCaw in Teledesic, which is working on an ambitious plan to employ hundreds of low-orbit satellites to provide a worldwide two-way broadband telecommunications service.

New Words and Phrases

architect ['ɑːkɪtekt] n. 建筑师；缔造者
worldwide ['wɜːl(d)waɪd] adj. 全世界的，世界范围的 adv. 在世界各地
Seattle [sɪ'ætl] n. 西雅图（美国一港市）
Harvard ['hɑːvəd] n. 哈佛大学；哈佛大学学生
hall [hɔːl] n. 过道，走廊；食堂；学生宿舍 n. （英）霍尔（人名）
executive [ɪɡ'zekjʊtɪv] adj. 行政的；经营的；执行的 n. 总经理；执行者
microcomputer ['maɪkrə(ʊ)kɒmˌpjuːtə] n. [计] 微型计算机
desktop ['desktɒp] n. 桌面；台式机
anticipation [æntɪsɪ'peɪʃ(ə)n] n. 希望；预感；先发制人；预支

vision   [ˈvɪʒ(ə)n]   n. 视力；美景；眼力；想象力 vt. 想象；显现；梦见
fundamentally   [fʌndəˈmentəlɪ]   adv. 根本地，从根本上；基础的
publish   [ˈpʌblɪʃ]   vt. 出版；发表；公布 vi. 出版；发行；刊印
critical   [ˈkrɪtɪk(ə)l]   adj. 鉴定的；爱挑剔的；决定性的；评论的
applause   [əˈplɔːz]   n. 欢呼，喝彩；鼓掌欢迎
biotechnology   [ˌbaɪə(ʊ)tekˈnɒlədʒɪ]   n. [生物] 生物技术；生物工艺学
visual   [ˈvɪʒjʊəl; -zj-]   adj. 视觉的，视力的；栩栩如生的
invest   [ɪnˈvest]   vt. 投资；覆盖；授予；包围 vi. 投资，入股；花钱买
cellular   [ˈseljʊlə]   adj. 细胞的；由细胞组成的 n. 移动电话；单元
ambitious   [æmˈbɪʃəs]   adj. 野心勃勃的；有雄心的；热望的；炫耀的
Harvard University   哈佛大学
Steve Ballmer   史蒂夫·鲍尔默
the MITS Altair   MITS Altair 计算机
thebest- seller lists   列入畅销书目录
the Road Ahead   未来之路；前方的道路
Amazon. com.   亚马逊网站
CraigMcCaw   克莱格·麦科考

Notes

[1] William (Bill) H. Gates is chairman and chief software architect of Microsoft Corporation, the worldwide leader in software, services and Internet technologies for personal and business computing.

译文：威廉〔比尔〕·H·盖茨是微软公司主席、首席软件设计师，也是全世界个人及商务计算机领域软件制作、服务、互联网技术的领军人物。

说明：这是一个主—系—表结构的简单句。其表语中心词为 chairman and chief software architect, of Microsoft Corporation 是后置定语；the worldwide leader 是表语中心词的同位语，in software, services... business computing 是介词短语作后置定语。

[2] Gates' anticipation and his vision for personal computing have been central to the success of Microsoft and the software industry.

译文：盖茨对个人计算机的远见卓识是微软公司和软件业成功的关键。

说明：这也是一个主—系—表结构的简单句。Gates' anticipation and his vision 是主语中心词，for personal computing 是后置定语，central（中心的，主要的）是形容词作表语，to the success... 是介词短语作状语，修饰谓语 have been central。

Exercises

Answer the following questions:

1. Where did Gates begin to have interest in software? And when did he start programming computers?
2. What did Gates develop for the first microcomputer—the MITS Altair at Harvard University?
3. With whom did Gates start to build Microsoft?
4. Whathave been central to the success of Microsoft and the software industry?
5. What did Gates also found besides Microsoft?

Chapter Ⅳ   Advanced Electronic & Communicative Engineering

Unit 18   Artificial Intelligence (AI)

Text

Since World War II, computer scientists have tried to develop techniques that would allow computers to act more like humans. The entire research effort, including decision-making systems, robotic devices (seen in Fig. 18-1), and various approaches to computer speech, is usually called artificial intelligence (AI).

An ultimate goal of AI research is to develop a computer system that can learn concepts (ideas) as well as facts, make commonsense decisions, and do some planning. In other words, the goal is to eventually create a "thinking, learning" computer.

A computer program is a set of instructions that enables a computer to process information and solve problems. Most programs are fairly rigid they tell the computer exactly what to do, step by step. AI programs are, however, exceptions to this rule. They can take short-cuts, make choices, search for and try out different solutions, and change their methods of operation.

Fig. 18-1   Robots

In many AI programs, facts are arranged to enable the computer to tell how many pieces of information relate to each other and to a given problem. "If/then" rules of reasoning are also programmed in to enable the computer to select, organize, and update its information. According to these rules, if something is true, then certain things must follow. [1] Every action makes new possible actions available.

Programs to play chess have been around since the early days of electronic computers, but they tended to be rigid and limited by the skills of the program designer. Detailed instructions on what moves to make and how to respond to an opponent's moves were written into a program. Sometimes the suggestions of several chess experts were included. However, such programs seldom defeated human chess experts. The computer program would tend to be strong in the opening part of the game, but would weaken as the game went on.

Thanks to AI research, all that has changed. Recently, chess-playing computer programs have been developed to defeat most human opponents—including chess masters.

Of course, there's more to artificial intelligence than the ability to play games. Computer scientists are working on dozens of different practical uses for AI programs. These include operating robots, solving math and science problems, understanding speech, and analyzing images.

Perhaps the biggest use of AI programs is expert advisors for trouble-shooting (locating problems and making repairs) complex systems ranging from diesel engines to nuclear submarines and to the human body. In other words, these AI programs search for trouble, detect and classify problem areas, and give advice.[2]

The use of AI expert advice systems will not be limited to trouble-shooting specific machinery. AI programs are being developed for economic planning, weather forecasting, casting, oil exploration, computer design, and numerous other uses.

AI techniques are also being used to analyze human speech and to synthesize speech.[3] With the help of laser sensors, AI techniques are being developed to analyze visual information and to improve robot capabilities.

Most artificial intelligence systems involve some sort of integrated technologies, for example the integration of speech synthesis technologies with that of speech recognition. The core idea of AI systems integration is making individual software components, such as speech synthesizers, interoperable with other components, such as common sense knowledge bases, in order to create larger, broader and more capable AI systems.[4] The main methods that have been proposed for integration are message routing, or communication protocols that the software components use to communicate with each other, often through a middleware blackboard system (seen in Fig. 18-2).

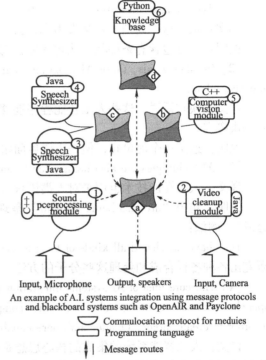

Fig. 18-2 Artificial Intelligence Integrated Systems

New Words and Phrases

entire [ɪnˈtaɪə; en-] adj. 全部的，整个的，全体的
ultimate [ˈʌltɪmət] adj. & n. 最后的，最终的，根本的，最终
commonsense [ˈkɔmənˈsɛns] adj. 常识的，具有常识的
techniques [tekˈniːks] n. 技巧，手法，技术，技能
detect [dɪˈtekt] vt. 察觉；发现；探测
approaches [əˈprəʊtʃɪz] n. 方法，近似(值)接近，走近
research [rɪˈsɜːtʃ; ˈriːsɜːtʃ] n. 追究，探测，调查，探索 vi. 做研究
rigid [ˈrɪdʒɪd] adj. 刚硬的，刚性的，严格的
advisor [ədˈvaɪzə] n. 顾问，指导教师，劝告者
short-cut [ˈʃɔːtkʌt] n. 捷径，近路 vi. 走捷径，抄近路
instructions [ɪnˈstrʌkʃənz] n. 教导，教诲 adj. 说明用法的，操作指南的
capabilities [ˌkeɪpəˈbɪlɪtɪs] n. 能力，可能，容量
synthesize [ˈsɪnθəsaɪz] vt. 合成；综合 vi. 合成；综合

synthesizer [ˈsɪnθəsaɪzɚ] *n.* 综合者，合成器
integration [ɪntɪˈgreɪʃ(ə)n] *n.* 结合，整合，一体化
trouble-shooting *n.* 故障寻找
diesel engine *n.* 柴油机
nuclear submarine *n.* 核潜艇

Notes

[1] According to these rules, if something is true, then certain things must follow.
译文：根据这一规则，若某事为真，另一些确定的事必随之发生。
说明：这是包含有条件状语从句的复合句，According to... 是介词短语作状语。

[2] In other words, these AI programs search for trouble, detect and classify problem areas, and give advice.
译文：换句话说，这些人工智能程序能够寻找故障、确定故障范围、对故障进行分类，并提供咨询。
说明：这是一个简单句，但是主语后面跟有几个并列谓语动词。

[3] AI techniques are also being used to analyze human speech and to synthesize speech.
译文：人工智能技术也用来对人类语言进行分析和合成。
说明：句中"are also being used to"是现在进行时被动语态；Synthesize 表示"综合；合成"，例如：
They have developed all kinds of tools to synthesize and manipulate this molecule. 他们已经研究出各种各样合成和处理这些分子的方法。

[4] The core idea of AI systems integration is making individual software components, such as speech synthesizers, interoperable with other components, such as common sense knowledge bases, in order to create larger, broader and more capable AI Systems
译文：人工智能系统的集成的核心思想是使单个软件组件，例如语音合成器，能够与其他组件诸如常识知识库进行交互操作，以创造更大、更广泛和更强的人工智能系统。
说明：这仍然是一个简单句。The core idea of AI systems integration 是主语，is making 是谓语，individual software components 是宾语；such as 引出同位语，对 software components 起补充说明作用；interoperable with... 是宾语补语；第二个 such as 也是引出同位语，对 other components 起补充说明作用；in order to... 是不定式短语作目的状语。

Exercises

I. Write True (T) or False (F) beside the following statements about the text.

1. Artificial intelligence includes only decision-making systems.
2. A computer program is a set of instructions that enables a computer to process information and solve problems.
3. Most programs are fairly flexible.
4. Artificial intelligence techniques are also being used to analyze human speech and to synthesize speech
5. The core idea of AI systems integration is making individual software components interoperable with other components.

Ⅱ. Match the following terms in column A with the appropriate definition or expression in column B.

Column A   Column B
1. decision-making   a. a person who gives advice
2. short-cut   b. the process of deciding about something important
3. advisor   c. a way of doing something that is quicker than the usual way

Ⅲ. Translate the following sentences into English.
1. 由于人工智能的出现，这一切都已经改变了。
2. 计算机程序是一组能让计算机对信息进行处理并做出决策的指令。
3. 人工智能技术也能用于人类语言的分析和合成。
4. 大部分人工智能系统都与某种集成技术有关。

Ⅳ. Translate the following paragraph into Chinese.

AI is a growing field that covers many disciplines. Subareas of AI include knowledge representation, learning, theorem proving, search, problem solving and planning, expert systems, natural-language (text or speech) understanding, computer vision, robotics, and several others (such as automatic programming, AI education, game playing, etc.). AI is the key for making technology adaptable to people. It will play a crucial role in the next generation of automated systems.

Useful Information

中外著名电子信息公司简介

作为即将跻身于现代企业的专业技术人员，应及时了解本行业国际国内知名公司的发展情况，努力缩小差距，保持与国际接轨，像诸多著名的中国品牌企业技术骨干一样，及早树立为中国赢得荣誉与地位的雄心。

1. 美国电话电报公司 www.att.com

美国电话电报公司（AT&T Corporation, AT&T）是一家美国电信公司，创建于1877年，曾长期垄断美国长途和本地电话市场。AT&T在近几十年中曾经过多次分拆和重组。目前，AT&T是美国最大的本地和长途电话公司，总部位于得克萨斯州圣安东尼奥。AT&T有8个主要部门：贝尔实验室、商业市场集团、数据系统公司、通用市场集团、网络运营集团、网络系统集团、技术系统集团、公司国际集团。主要业务：①为国内国际提供电话服务。利用海底电缆、海底光缆、通信卫星，可联系250个国家和地区，147个国家和地区可直接拨号。②提供商业机器、数据类产品和消费类产品。③提供电信网络系统。④各种服务及租赁业务。它还为政府提供产品和服务。该公司非常重视科研和开发新产品。

2. 西门子 www.siemens.com

总部位于柏林和慕尼黑的西门子集团公司是世界上较大的电气工程和电子公司之一。自公司成立以来，可持续性就一直是西门子公司的显著特征。西门子是一家大型国际公司，其业务遍及全球190多个国家，在全世界拥有大约600家工厂、研发中心和销售办事处。公司的业务主要集中于6大领域：信息和通信、自动化和控制、电力、交通、医疗系统和照明。

3. 三星 www.samsung.com

三星电子（Samsung Electronics）是三星集团子公司之一。三星电子的主要经营项目有5项：通信（手机和网络）、数字式用具、数字式媒介、液晶显示器和半导体。

4. 荷兰皇家飞利浦电子公司 www.philips.com

荷兰皇家飞利浦电子公司，简称为飞利浦，是世界上较大的电子公司之一。产品主要涉及照明、家庭电器、医疗等系统。总部位于荷兰的飞利浦公司在全球 60 多个国家拥有大约 128000 名员工。公司在心脏监护、紧急护理与家庭医疗保健、节能照明解决方案与新型照明应用方面，以及针对个人舒适优质生活的平板电视、男性剃须和仪容产品、便携式娱乐产品及口腔护理产品等领域均居于世界较高地位。

5. 海尔 www.haier.com

海尔集团是世界白色家电著名品牌。海尔在全球建立了 29 个制造基地、8 个综合研发中心、19 个海外贸易公司，全球员工总数超过 6 万人，已发展成为大规模的跨国企业集团。海尔品牌旗下的电冰箱、空调、洗衣机、电视机、热水器、计算机、手机和家居集成等 19 个产品被评为中国名牌，其中海尔电冰箱、洗衣机还被原国家质检总局评为首批中国世界名牌。

6. 海信集团有限公司 www.hisense.com

海信集团是特大型电子信息产业集团公司，成立于 1969 年。如今，已形成了以数字多媒体技术、现代通信技术和智能信息系统技术为支撑，涵盖多媒体、家用电器、通信、智能信息系统和现代地产与服务的产业格局。

海信拥有海信电器和海信科龙电器两家在沪、深、港三地的上市公司，同时成为国内唯一一家持有海信（Hisense）、科龙（Kelon）和容声（Ronshen）三个中国驰名商标的企业集团。海信电视、海信空调、海信电冰箱、海信手机、科龙空调、容声电冰箱全部当选为中国名牌，海信电视、海信空调、海信电冰箱全部被评为国家免检产品，海信电视首批获得国家出口免检资格。

7. 华为技术有限公司 www.huawei.com.cn

华为技术有限公司是一家总部位于中国广东省深圳市的生产销售通信设备的员工持股的私营通信科技公司。华为的产品主要涉及通信网络中的交换网络、传输网络、无线及有线固定接入网络和数据通信网络及无线终端产品，为世界各地通信运营商及专业网络拥有者提供硬件设备、软件、服务和解决方案。

8. 方正集团有限公司 www.founderpku.com

方正集团由北京大学 1986 年投资创办，北京大学持股 70%，管理层持股 30%。依托北京大学，方正拥有并创造对中国 IT、医疗医药产业发展至关重要的核心技术；开放、规范的资本平台吸引了如英特尔、欧姆龙、瑞士信贷、东亚银行在内的国际资本注入。方正已快速成长为综合实力与华为、海尔同列中国信息产业前列的大型控股集团公司。

Reading

Computer Vision

Computer vision means artificial sight by means of computer and other pertinent techniques. Although today's computer vision systems are crude, compared with human sight, they are promising development of computer science and technology and will have brilliant prospects.

A computer mimics human sight in four steps: image acquisition, image processing, image analysis and image understanding (seen in Fig. 18-3).

1. **Image Acquisition**

A TV camera is usually used to take instantaneous images and transform them into electrical sig-

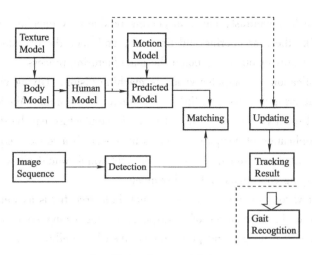

Fig. 18-3 Computer Vision

nals, which will be further translated into binary numbers for the computer to handle. The TV camera scans one line at a time. Each line is further divided into hundreds of pixels. The whole frame is divided into hundreds (for example, 625) of lines. The brightness of a pixel can be represented by a binary number with certain bits, for example, 8 bits. The value of the binary number varies from 0 to 255 ($=2^8-1$), a range great enough to accommodate all possible contrast levels of images taken from real scenes. These binary numbers are stored in an RAM (it must have a great capacity) ready for further processing by the computer.

2. Image Processing

Image processing is for improving the quality of the images obtained. First, it is necessary to improve the signal-to-noise ratio. Here noise refers to any interference flaw or aberration that obscure the objects on the image. Second, it is possible to improve contrast. Enhance sharpness of edges between images through various computational means.

3. Image Analysis

It is for outlining all possible objects that are included in the scene. A computer program checks through the binary visual information in store for it and identifies specific feature and characteristics of those objects. Edges or boundaries are identifiable because of the different brightness levels on either side of them. Using certain algorithms, the computer program can outline all possible boundaries of the objects in the scene. Image analysis also looks for textures and shadings between lines.

4. Image Comprehension

Image comprehension means understanding what is in a scene. Matching the presorted binary visual information with certain templates which represent specific objects in a binary form is a technique borrowed from artificial intelligence, commonly referred to as "template matching". One by one, the templates are checked against the binary information representing the scene. Once a match occurs, an object is identified. The template matching process continues until all possible objects in the scene have been identified, otherwise it fails.

Computer vision has found its way into industries, doing jobs which used to be exclusive for human operators, for example, identifying specific objects or patterns by finding out distinctive details in shape, size, color, contrast, etc., inspecting products by checking for any flaws such as crack, smear. Jobs like these are boring and tiresome, and once the operator gets bored or tired, he/she tends to miss details or otherwise botch the manufacturing process.

One popular application of computer vision is machine vision. Giving vision to machines can automate their manufacturing process with little or no human intervention. Machine vision greatly improve productivity and quality, and the cost and time of manufacture can be dramatically reduced.

The promising application of computer vision is in robots. If a robot is equipped with computer vision, it becomes an intelligent robot. A robot with sight which would allow it to adjust its operations to fit itself to varying conditions and environments.

Today, computer vision is extensively used in non-industrial fields as well, for example, for identifying fingerprints or facial features of a suspect, distinguishing counterfeit notes and forged paintings, analyzing medical images and photographs taken by satellites, etc.

New Words and Phrases

pertinent	[ˈpɜːtɪnənt]	adj.	相关的，相干的，中肯的，切题的
manufacture	[mænjʊˈfæktʃə]	vt.	制造，生产　n. 制成品，产品，工业
promising	[ˈprɒmɪsɪŋ]	adj.	有希望的，有前途的　v. 许诺，答应
productivity	[ˌprɒdʌkˈtɪvɪtɪ]	n.	生产率，生产力，生产能力
prestore	[priːˈstɔː]	v.	预存储
instantaneous	[ˌɪnst(ə)nˈteɪnɪəs]	adj.	瞬间的，即刻的，猝发的
templet	[ˈtemplɪt]	n.	样板，模板（等于 template）
distinguishing	[dɪˈstɪŋgwɪʃɪŋ]	adj.	有区别的　v. 区别，表现突出
counterfeit	[ˈkaʊntəfɪt; -fiːt]	vt.	假装，伪装　n. 伪造品　adj. 伪造的
accommodate	[əˈkɒmədeɪt]	vt.	供应，使适应，容纳　vi. 适应，调解
distinctive	[dɪˈstɪŋ(k)tɪv]	adj.	有特色的，与众不同的
brightness	[ˈbraɪtnɪs]	n.	亮度，明亮，光泽度，灯火通明，活泼，愉快
dramatically	[drəˈmætɪkəlɪ]	adv.	引人注目地，戏剧地，显著地，剧烈地
prospect	[ˈprɒspekt]	n.	前景，期望，眺望处，景象　vt. 勘探，勘察
algorithm	[ˈælgərɪð(ə)m]	n.	运算法则，演算法，计算程序

Exercises

Ⅰ. Fill in the blank without referring to the original text. Then check your answers against the original. After that, read the passage aloud until you can say it by your memory.

One popular application of computer vision is _____ vision. Giving vision to machines can _____ their manufacturing process with little or no _____ intervention. Machine vision greatly improve productivity and _____, and the cost and time of manufacture can be _____ reduced.

Ⅱ. Translate the following sentences into Chinese.

1. Image analysis also looks for textures and shadings between lines.

2. The template matching process continues until all possible objects in the scene have been identified, otherwise it fails.

3. If a robot is equipped with computer vision, it becomes an intelligent robot.

Unit 19 Sensor Technology

Text

A sensor is a device, which responds to an input quantity by generating a functionally related output usually in the form of an electrical or optical signal. [1]

Sensors and sensor systems perform a diversity of sensing functions allowing the acquisition, capture, communication, processing and distribution of information about the states of physical systems. This may be chemical composition, texture and morphology, large-scale structure, position, and so on. Few products and services of the modern society would be possible without sensors.

1. The Sensor Value Chain

Sensor technology is distinctly interdisciplinary. Few organizations have all the competencies necessary for the realization of a sensor solution in-house.

The realization of a sensor product requires tasks to be completed, ranging all the way from product definition to final product and subsequent marketing and service. The Fig. 19-1 shows this value chain of sensor development and the main application areas.

- Chemical engineering industry.
- Metal & plastic processing.
- Wood, textile etc.
- Transport.
- Domestic appliances & entertainment.
- Energy.
- Environment.
- Health care.
- Food processing.
- Agriculture.
- Security & defense.

2. Wireless Sensor Technology

For applications within health care, industrial automation, consumer products and security there is a strong and growing need for wireless, self-powered sensors. Radio frequency identification technology (RFID) is an example of an emerging application with great potential. Sensors with wireless connections and no internal power supply are anticipated to become of great importance in areas like health care, consumer products, and structural health monitoring, Energy tapping, i. e.

The Fig. 19-2 illustrates a sensor field where a large number of connected sensor nodes are embedded. Each node will consist of a wireless sensor often without any internal power supply. The sensor interacts with a transceiver which is again connected to an infrastructure, possibly to a so-called sink. The collection of data may be controlled by a managing device.

3. Biometric Sensors

This is another area where some markets are expected to exhibit strong growth over the next years. Fingerprint identification equipment and iris scanners are examples of such markets that are

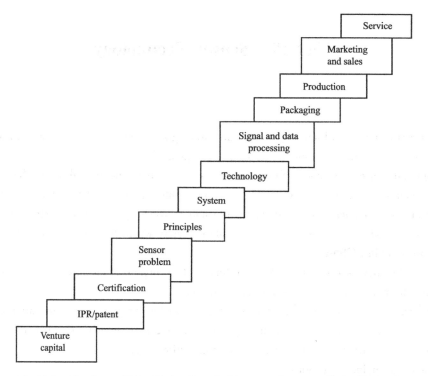

Fig. 19-1 The Sensor Value Chain: Rom Problem to a Commercially Viable Product

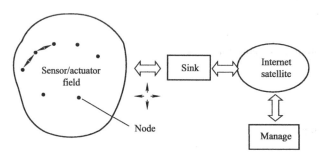

Fig. 19-2 Wireless Sensor Network

spurred by the increasing demand within security.

4. Non-invasive & Non-contact Sensors

An increasing number of applications call for non-contact sensing. Light and sound play important roles both independently and combined. Here, combinations of ultrasound and light are expected to become important in order to overcome the limitations inherent when using either light or sound independently.[2]

5. Miniaturization and Integration

Sensors are often used in large production plants. However, on-line sensing often has to be done in areas with limited space. Hostile environments require robust sensors and robustness may be obtained by miniaturization and integration. In optics, this may imply new system and fibre optics. In acoustics, new non-contact methods for excitation are being devised.

6. Novel Materials

Novel materials are needed if a number of very different functional requirements are to be fulfilled at sufficiently low cost. It may be systems combining microfluidics with light generation and detection. Polymers are anticipated to play an increasingly larger role, both due to potential low cost and due to great flexibility in functional properties. Recent development of advanced microscopes has made it possible to see and even move single atoms and molecules. This opens opportunities for creating entirely new materials and processes. The technology has become known as nanotechnology and currently receives a lot of attention.

Not many companies have thus far reached the stage of commercialization, but intense research is in progress and expectations are very high.

7. Sensor Fusion and Sensor Networks

Complex systems will often be monitored by a number of very different types of sensors. X-ray can reveal properties of weldings on-line, optics can detect chemical composition and macroscopic dynamics, whereas ultrasound may provide information about the inner structures of systems.

Sensor fusion is combining such multisensory information in order to obtain new functionality. Moreover, systems with a large number of low-cost sensors coupled in networks are becoming increasingly important.

New Words and Phrases

capture ['kæptʃə] vt. 俘获，夺得 n. 战利品，俘虏
texture ['tekstʃə] n. 质地，肌理，结构，本质，实质
anticipated [æn'tɪsəˌpeɪtɪd] adj. 预期的，期望的 v. 预料，盼望
morphology [mɔː'fɒlədʒɪ] n. [生物] 形态学、形态论
interdisciplinary [ˌɪntə'dɪsɪplɪn(ə)rɪ] adj. 各学科间的
subsequent ['sʌbsɪkw(ə)nt] adj. 后来的，随后的
vibration [vaɪ'breɪʃ(ə)n] n. 振动，颤动，摇动，摆动，犹豫，心灵感应
monitoring ['mɒnɪtərɪŋ] n. 监视，[自] 监控，检验 v. 监视，监听，监督
embedded [ɪm'bedɪd] v. 嵌入 adj. 嵌入式的，植入的，内含的
transceiver [træn'siːvə; trɑːn-] n. 无线电收发机，收发器
ultrasound ['ʌltrəsaʊnd] n. 超频率音响，超声波
inherent [ɪn'hɪər(ə)nt; -'her(ə)nt] adj. 固有的，内在的，与生俱来的
hostile ['hɒstaɪl] n. 敌对 adj. 敌对的，敌方的，怀敌意的
spur [spɜː] n. 鼓舞，刺激 vi. 骑马疾驰；给予刺激 vt. 激励，鞭策
acoustics [ə'kuːstɪks] n. 声学，音响效果，音质
robust [rə(ʊ)'bʌst] adj. 强健的，健康的，粗野的，粗鲁
miniaturization [ˌmɪnɪətʃərɪˈzeɪʃən] n. 小型化，微型化
polymer ['pɒlɪmə] n. 聚合物
macroscopic [ˌmækrə(ʊ)'skɒpɪk] adj. 宏观的，肉眼可见的
multisensory [ˌmʌltɪ'sensərɪ] adj. 多种感觉的；使用多种感觉器官的
nanotechnology [ˌnænə(ʊ)tek'nɒlədʒɪ] n. 纳米技术
self-powered 自供电的
RFID 射频识别技术

fingerprint identification 指纹识别
chemical composition 化学成分，化学组成
iris scanner 虹膜扫描仪
non-invasive 非插入式
novel material 新型材料

Notes to the text

[1] Sensors with wireless connections and no internal power supply are anticipated to become of great importance in areas like health care, consumer products, and structural health monitoring, Energy tapping, i. e.

译文：无线连接和没有内部电源的传感器预期将在卫生保健、消费品、结构健康监测和能源发掘等领域发挥重要的作用。

说明：这是一个简单句。句中"with wireless connections... supply"是带有 with 的复合结构作后置定语，修饰主语中心词"sensors"；are anticipated to become of 是复合结构作谓语；great importance 作宾语；in areas 是介词短语作状语；like health care... 是介词短语作后置定语修饰 areas。

[2] Fingerprint identification equipment and iris scanners are examples of such markets that are spurred by the increasing demand within security.

译文：指纹识别设备和虹膜扫描设备就是实例，它们是由人们不断提高的安全需求而发展起来的。

说明：这是一个复合句。其中 that 引导的定语从句修饰先行词 examples，主句中的先行词与定语从句之间被 of such markets 这样一个关系更紧密的后置定语隔开了。

[3] Hostile environments require robust sensors and robustness may be obtained by miniaturization and integration.

译文：恶劣的环境需要优质的传感器，而其优良性能可以通过小型化和集成化来得到。

说明：本句是并列句，and 是并列连词，在这里表示两类事物的对比与对照。

[4] Polymers are anticipated to play an increasingly larger role, both due to potential low cost and due to great flexibility in functional properties.

译文：高分子聚合物预期将在这个领域扮演越来越重要的角色，不仅是因为它成本低，还由于它功能特性上具有很高的柔韧性。

说明：这是一个简单句。主语是 Polymers，are anticipated to play 是复合结构作谓语，an increasingly larger role 作宾语，both due to... and due to... 是并列成分作原因状语。

[5] X-ray can reveal properties of welding on-line, optics can detect chemical composition and macroscopic dynamics, whereas ultrasound may provide information about the inner structures of systems.

译文：X 射线可以在线显示焊接的质量，光可以观测化学组成和宏观动力学，而超声波能提供系统内部构造信息。

说明：本句是并列句，三个分句都是简单句。句中"whereas"连词，意思是"但是，然而，鉴于"，用以比较或对比两种或两种以上的事物特性。

Exercises

Ⅰ. Translate the following phrases and words into Chinese.

1. ultrasonic
2. energy tapping
3. fingerprint identification
4. microscopic
5. macroscopic
6. nanotechnology technology
7. health care
8. consumer product

Ⅱ. Decide whether the following statements are true (T) or False (F) according to the text.

1. Few products and services of the modern society would be possible without sensors.

2. Polymers are anticipated to play an increasingly important role, mainly due to potential low cost.

3. many companies have thus far reached the stage of commercialization.

4. Biometric sensors is another area where some markets are expected to exhibit strong growth over the next years.

5. X-ray can detectchemical composition and macroscopic dynamics, optics can reveal properties of weldings on-line, whereas ultrasound may provide information about the inner structures of systems.

Ⅲ. Translate the following passage into Chinese.

New materials are needed if a number of very different functional requirements are to be fulfilled at sufficiently low cost. It may be systems combining microfluidics with light generation and detection. Polymers are anticipated to play an increasingly larger role, both due to potential low cost and due to great flexibility in functional properties. Recent development of advanced microscopes has made it possible to see and even move single atoms and molecules. This opens opportunities for creating entirely new materials and processes. The technology has become known as nanotechnology and currently receives a lot of attention.

Practical English

怎样阅读英文招聘广告

英语招聘广告是广告中的一种，属于非正式文体。英语招聘广告中的资格要求和工作职责通常是一条条列出来的，简洁明了、清晰明确、引人注意。

1. 注意缩略语

（1）首字母缩略词（Acronyms）

首字母缩略词指保留每个单词的第一个字母，而把后面的字母省略，多用于表示国家、地区和机构的专有名词、表示学位的名词和一些习惯搭配的名词。

例如：GE = General Electric Corporation（通用电气公司），ECM = European Common Market（欧洲共同体），MBA = Master of Business Administration（工商管理硕士）。

（2）去尾缩略词（Words Shortened via Back-clipping）

这类缩略词把一个单词的后半部分去掉，只保留前面的2~5个字母。例如：knowl = knowledge（知识），loc = location（位置、场所），inexp = inexperienced（无经验的），pref = preference（优先），corp = corporation（公司）。

（3）去元音缩略词（Abbreviations Formed by Omitting Vowels）

这类词把单词中的元音字母去掉，只剩下辅音字母。例如：hr = hour（小时），wk = week/work（周/工作），yr（s）= year（s）（年），mst = must（必须）。

（4）混合缩略词（Blended Abbreviations）

这类词的构成包括以下四种情况：

1）保留词的开头几个字母和最后一两个字母，如 agcy = agency（经商），appt = appointment（职务），asst = assistant（助理），oppty = opportunity（机会）等。

2）把所有元音字母去掉，同进又根据需要把中间的一两个辅音字母也去掉，如 hqtrs = headquarters（总部），stmts = statements（报告）等。

3）只保留词的开关和末尾各一个字母，如 Jr = junior（初级），Sr = senior（高级），gd = good（好）。

4）带'的，如 int'l = international（国际性的）等。

2. 省略句的理解

（1）省略主语

招聘广告对工作内容和性质、工作职责做出说明时通常省略主语。例如，

1）Responsibilities：Competitive intelligence：devise structure, practice, database and system of competitive intelligence in Asia with special on greater China.

2）Requirements：Bachelor's degree or above in sciences; 3 years of experience in analysis.

（2）省略动词

省略的动词通常包括系动词"be"、实义动词"have"，助动词"do"。

1）省略系动词 be。例如，Willing to learn and enter into new working environment.

2）省略实义动词 have。例如，At least 3 years relevant working experience in foreign companies.

3）省略助动词 do 或 does，这一情况不太多。例如，IBM Not Always Have Vacancies.

4）省略情态动词 must 或 can。例如，Work under pressure.（work 前省略了"can"一词）Be able to lift 35 pounds.（系动词 be 前省略了情态动词"must"）

（3）省略冠词

英语招聘广告是非正式文体，所以很多情况下冠词都被省略。例如，Bachelor degree and above; Good command of English.

3. 英文招聘广告范例

Marketing Assistant

Responsibility：

- Responsible for the local management of marketing and sales activities according to instructions from head office.
- Collect related information for the head office.

Requirements：

- College degree or above with good English (speaking and writing).
- Develop relationship with local media and customers.
- With basic idea of sales and marketing, related experience is preferred.
- Working experience in an international organization is a must.
- Good communication and presentation skills.

Reading

Remote Sensing

Remote sensing is the process of collecting data about objects or landscape features without coming into direct physical contact with them. Most remote sensing is performed from aircraft or satellites using instruments, which measure electromagnetic radiation (EMR) that is reflected or emitted from the terrain (as shown in Fig. 19-3). In other words, remote sensing is the detection and measurement of electromagnetic energy emanating from distant objects made of various materials. This is done so that we can identify and categorize these objects by class or type, substance, and spatial distribution.

Remote sensing devices can be differentiated in terms of whether they are active or passive. Active systems, such as radar and sonar, beam artificially produced energy to a target and record the reflected component. Passive systems, including the photographic camera, detect only energy emanating naturally from an object, such as reflected sunlight or thermal infrared emissions. Today, remote sensors, excluding sonar devices, are typically carried on aircraft and earth-orbiting spacecraft.

To complete the remote sensing process, the data captured and recorded by remote sensing systems must be analyzed by interpretive and measurement techniques in order to provide useful information about the subjects of investigation. These techniques are diverse, ranging from traditional methods of visual interpretation to methods using sophisticated computer processing. Accordingly, the two major components of remote sensing are data capture and data analysis. [1]

Fig. 19-3 Remote Sensing

Thus today we find there are two major branches of remote sensing. The branch first mentioned above is referred to as "image-oriented" because it capitalizes on the pictorial aspects of the data and utilizes analysis methods which rely heavily on the generation of an image. The second branch is referred to as "numerical-oriented" because it results directly from the development of the computer and because it emphasizes the quantitative aspects of the data, treating the data abstractly as a collection of measurement. In this case an image is not thought of as data but rather as a convenient mechanism for viewing the data.

Today we are acquiring earth observational data from earth-orbiting satellites, because of the wide view possible from satellite altitudes, the speed with which the satellite borne sensors travel, and the number of spectral bands used, very large quantities of data are being produced.

Satellite remote sensing may be done two ways:
- Using passive sensor systems—Contains an array of small detectors or sensors that can detect electro-magnetic radiations emitted from the earth's surface.
- Using active sensor systems—The system sends out electromagnetic radiation towards target object (s) and measures the intensity of the return signal. [2]

Data collected by the satellites are then transmitted to ground stations wherein images of earth's surface are reconstituted to obtain the required information. [3]

Take a look at few benefits of satellite remote sensing:
- Enables continuous acquisition of data;
- Helps to receive up-to-date information (satellite remote sensing can be programmed to enable regular revisit to object or area under study);
- Offers wide regional coverage and good spectral resolution;
- Offers accurate data for information and analysis.

New Words and Phrases

landscape	[ˈlæn(d)skeɪp]	n. 风景，山水画，地形，美化
reflect	[rɪˈflekt]	vt. 反映，反射，表达，显示 vi. 映现，深思
emanate	[ˈemaneɪt]	vi. 发出，散发，发源 vt. 放射，发散
categorize	[ˈkætəgəraɪz]	vt. 分类
distribute	[dɪˈstrɪbjuːt; ˈdɪstrɪbjuːt]	vt. 分配，散布，分开，把……分类
reflect	[rɪˈflekt]	v. 反射（光、热、声或影像）
sophisticate	[səˈfɪstɪkeɪt]	n. 久经世故者，精通者 vt. 弄复杂 vi. 诡辩
sophisticated	[səˈfɪstɪkeɪtɪd]	adj. 复杂的，精致的；久经世故的
diverse	[daɪˈvɜːs; ˈdaɪvɜːs]	adj. 多种多样的，不同的；变化多的
thermal	[ˈθɜːm(ə)l]	adj. 热的，保热的；温热的 n. 上升的热气流
array	[əˈreɪ]	n. 队列，阵列，数组，排列，大批，一系列，衣服 vt. 部署
intensity	[ɪnˈtensɪtɪ]	n. 强烈，强度，强烈；[电子] 亮度；紧张
reconstitute	[riːˈkɒnstɪtjuːt]	vt. 再组成，再构成；重新设立
capitalize	[ˈkæpɪtəlaɪz]	vi. 利用，积累资本 vt. 使资本化；以大写字母写
mechanism	[ˈmek(ə)nɪz(ə)m]	n. 方法，途径，程序，机械装置，原理
electromagnetic	[ɪˌlektrə(ʊ)mægˈnetɪk]	adj. 电磁的
investigation	[ɪnˌvestɪˈgeɪʃ(ə)n]	n. 调查，科学研究，学术研究调查
infrared	[ɪnfrəˈred]	adj. 红外线的；n. 红外线

Notes

[1] Contains an array of small detectors or sensors that can detect electro-magnetic radiations emitted from the earth's surface.

译文：它包括一排小型探测器或传感器阵列，可以检测从地球表面发射的电磁辐射。

说明：本句是复合句。That 引导的定语从句修饰 detectors or sensors。

[2] The system sends out electromagnetic radiation towards target object (s) and measures the intensity of the return signal.

译文：系统向一个或多个目标对象发出电磁辐射，并测量回波信号的强度。

说明：本句是简单句，包含有两个并列的谓语动词 sends out 和 measures。

[3] Data collected by the satellites are then transmitted to ground stations where images of earth's surface are reconstituted to obtain the required information.

译文：由卫星收集的数据传输到地面站，在那里，地球表面的图像将被重组以获取所需的信息。

说明：本句是复合句，where 引导的定语从句修饰 ground stations。

Exercises

I. Fill in the missing words according to the text

1. Most remote sensing is performed from aircraft or satellites using _____.

2. Using _____ sensor systems— Contains an array of small detectors or sensors that can detect electro-magnetic radiations _____ from the earth's surface.

3. Data collected by the satellites are then _____ to ground stations wherein images of earth's surface are _____ to obtain the required information.

II. Translate the following sentences into Chinese.

1. Even if you already know that a string is an array of characters, read this part.

2. When the thief used the server, his files were transmitted to the backup site.

3. Until then, however, he is working with reconstituted silkworm silk, making novel films and other materials.

Unit 20 Internet of Things

Text

Internet of Things (IOT), also known as the sensor network, refers to a variety of information sensing devices and the Internet combine to form a huge network, will enable all of the items and network connections to facilitate the identification and management. [1] Because of its comprehensive sense, reliable delivery, intelligent processing features, it is considered as another wave of the information industry after the computer, the internet and mobile communication network.

Touch of a button on the computer or cell phone, even thousands of miles away, you can learn the status of an item, a person's activities. Send a text message, you can turn on the fan; if an illegal invasion of your home takes place, you will receive automatic telephone alarm. They are not just the scenes in Hollywood sci-fi blockbusters. They are gradually approaching in our lives.

It can be achieved due to the "things" in which there is a key technology for information storage object called radio frequency identification (RFID). [2] An RFID system consists of three components (as shown in Fig. 20-1): an antenna transceiver (often combined into a reader) and a transponder (the tag). The antenna emits radio signals to activate the tag and to read and write data to it. When activated, the tag transmits data back to the antenna. The data transmitted by the tag may provide identification or location information, or specifics about the product tagged, such as price, color, date of purchase, etc. Low-frequency RFID systems (30 kHz to 500 kHz) have short transmission ranges (generally less than six feet). High-frequency RFID systems (850 MHz to 950 MHz and 2.4 GHz to 2.5 GHz) offer longer transmission ranges (more than 90 feet). In general, the higher the frequency, the more expensive the system.

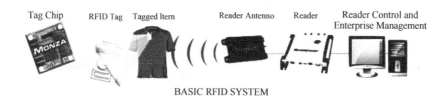

BASIC RFID SYSTEM

Fig. 20-1 An RFID System

For example, in mobile phones, embedded RFID-SIM card, your phone "information sensing device" can be connected with the mobile network. This phone can not only confirm the user's identity but also to pay the bills for water, gas and electricity, lottery, airline tickets and other payment services.

As long as an object embedded in a specific radio frequency tags, sensors and other devices connected to the Internet will be able to form a large network systems. [3] On this line, even thousands of miles away, people can easily learn and control the information of the object (as shown in Fig. 20-2).

To speak more concretely, let's imagine a world in which a large number of things that surround us are "autonomous", because they have:
- a name: a tag with a unique code;
- a memory: to store everything that they cannot obtain immediately from the net;
- a means of communication: mobile and energy-efficient, if possible;
- sensors: in order to interact with their environment;
- acquired or innate behaviors: to act according to logic, an objective given by its owner.

And, of course, like everything else on Earth, these things must have an electronic existence on the network.

Some experts predict that, in 10 years, "things" may become very popular, and develop into a trillion-scale high-tech market. Then, in almost all areas, such as the personal health, traffic control, environmental protection, public safety, industrial monitoring, elderly care, "things" will play a role. Some experts said that in only three to five years' time, it will change people's way of life.

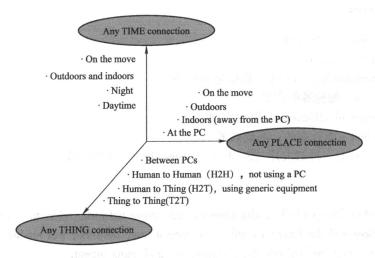

Fig. 20-2 How will Internet of Things Affect our Lives?

The Internet of Things has great promise, yet business, policy, and technical challenges must be tackled before these systems are widely embraced. Early adopters will need to prove that the new sensor driven business models create superior value. Industry groups and government regulators should study rules on data privacy and data security, particularly for uses that touch on sensitive consumer information. On the technology side, the cost of sensors and actuators must fall to levels that will spark widespread use. Networking technologies and the standards that support them must evolve to the point where data can flow freely among sensors, computers and actuators. Software to aggregate and analyze data, as well as graphic display techniques, must improve to the point where huge volumes of data can be absorbed by human. [4]

New Words and Phrases

  sensor ['sensə] *n.* 传感器
  facilitate [fə'sɪlɪteɪt] *vt.* 使容易，促进，帮助
  identification [aɪˌdentɪfɪ'keɪʃ(ə)n] *n.* 鉴定，识别；认同；身份证明

comprehensive　　[ˌkɒmprɪˈhensɪv]　　adj. 综合的；广泛的；有理解力的
delivery　　[dɪˈlɪv(ə)rɪ]　　n. 交付；分娩；递送
illegal　　[ɪˈliːgl]　　adj. 非法的；违法的；违反规则的　n. 非法移民；间谍
transponder　　[trænˈspɒndə]　　n. 应答器；转调器，变换器
tag　　[tæg]　　n. 标签；vt. 尾随；连接；添饰　vi. 紧随
embed　　[ɪmˈbed; em-]　　vt. 栽种；使嵌入，使插入；使深留脑中
autonomous　　[ɔːˈtɒnəməs]　　adj. 自治的；自主的；自发的
innate　　[ɪˈneɪt; ˈɪneɪt]　　adj. 先天的；固有的；与生俱来的
interact　　[ˌɪntərˈækt]　　v. 交互
embrace　　[ɪmˈbreɪs; em-]　　v. 拥抱，包括
actuator　　[ˈæktjʊeɪtə]　　n. 执行机构；激励者；促动器
spark　　[spɑːk]　　n. 火花；闪光　vt. 发动；鼓舞　vi. 闪烁；发火花
aggregate　　[ˈægrɪgət]　　vt. 使聚集，使积聚；总计达　n. 合计　adj. 总机的

Technical Terms

Internet of Things　　物联网
sensor network　　传感网
mobile communication network　　移动通信网络
product tagged　　加标签的产品
radio frequency identification　　射频识别
energy-efficient　　能效高的；高能效的
sci-fi blockbusters　　科幻大片（Sci-fi 是 Science-fiction 的缩写形式）

Notes

[1] Internet of Things (IOT), also known as the sensor network, refers to a variety of information sensing devices and the Internet combine to form a huge network, will enable all of the items and network connections to facilitate the identification and management.

译文：物联网也称为传感网，是指将各种信息传感设备与互联网结合起来而形成的一个巨大网络，这个网络可使所有的物品与其相连，方便身份识别和管理。

说明：本句是复合句。句子中 also known as the sensor network 是主句主语 Internet of Things 的同位语，而主句谓语是 refers to 和 will enable；a variety of ... network 为宾语从句作 refers to 的宾语；to form a huge network 与 to facilitate the identification and management 是不定式短语作目的状语；"to facilitate" 意思是 "使容易，促进，帮助"，例如，To facilitate the measurements, Professor Gross and his team captured the photon in a special box, a resonator. 为了方便测量，格罗斯教授和他的团队通过一个特殊的盒子——谐振器，来捕获光子。

[2] It can be achieved due to the "things" in which there is a key technology for information storage object called radio frequency identification (RFID).

译文：实现这一切是因为物联网里有一个存储物体信息的关键技术叫射频识别（RFID）。

说明：RFID 是 radio frequency identification 的缩写形式，即射频识别，又称为无线射频识别，是一种通信技术，可通过无线电信号识别特定目标并读写相关数据，识别工作无须人工干预，无须识别系统与特定目标之间建立机械或光学接触。最常见的应用有门禁控制和汽车的 ETC 收费等。

[3] As long as an object embedded in a specific radio frequency tags, sensors and other devices connected to the Internet will be able to form a large network systems.

译文：只要将特定物体嵌入射频标签，传感器和其他设备与互联网连接后，就能够组成一个庞大的网络系统。

说明：这是一个复合句，句中 As long as 引导条件状语从句，sensors and other devices 为句子主句的主语，connected to the Internet 为过去分词短语作定语修饰主语。

[4] Software to aggregate and analyze data, as well as graphic display techniques, must improve to the point where huge volumes of data can be absorbed by human.

译文：用来收集和分析数据的软件，以及图形显示技术，必须改进提高到这样一种地步——海量的数据能够被人接收。

说明：这是一个复合句，句中 Software to aggregate and analyze data 为主语，as well as 引导并列主语 graphic display techniques；must improve to 为谓语，where 引导定语从句修饰宾语 the point。improve 的意思是"改善，增进"例如，China is in a transition from its planned economy to a socialist market economy. Many of its current systems remain to be improved. 中国正处于计划经济体制向社会主义市场经济转变的时期，各方面的制度还不完善。

Exercises

I. Answer the following questions according to the text.

1. What is Internet of Things (IOT)?
2. Internet of Things is considered a wave of information industry; Can you mention other so called waves of information industry according to the text?
3. How many key components a RFID consist of and what are them?
4. What is the relation between thetransmission range and frequency of a RFID system?

II. Match the following phrases in column A with column B.

Column A	Column B
1. Internet of Things	a. 传感网
2. mobile communication network	b. 执行机构
3. sensor network	c. 能效高的
4. radio frequency identification	d. 射频识别
5. energy-efficient	e. 标签
6. actuator	f. 应答器
7. transponder	g. 移动通信网络
8. tag	h. 物联网

III. Fill in the blanks with the proper word. Change the form if necessary.

physical	across	collect	creating
integration	enable	defined	embedded

The internet of things (IoT) is the internetworking of _____ devices, vehicles, buildings and other items-_____ with electronics, software, sensors, actuators, and network connectivity that _____ these objects to _____ and exchange data. In 2013 the Global Standards Initiative on Internet of Things (IoT-GSI) _____ the IoT as "the infrastructure of the information society." The IoT allows objects to be sensed and/or controlled remotely _____ existing network infrastructure, _____ opportunities for more direct _____ of the physical world into computer-based systems, and resulting in improved efficiency, accuracy and economic benefit.

Ⅳ. Translate the following sentences into Chinese.

1. An RFID system consists of three components: an antenna transceiver (often combined into a reader) and a transponder (the tag).

2. As long as an object embedded in a specific radio frequency tags, sensors and other devices connected to the Internet will be able to form a large network systems.

3. The ability to network embedded devices with limited CPU, memory and power resources means that IoT finds applications in nearly every field.

Practical English

怎样用英文写个人简历

一般英文简历主要包括个人资料、教育背景、工作经历和兴趣爱好4部分。

1. 个人资料（Personal Data）

（1）名字

以"李扬"为例，标准的英文名写法是 Yang Li。双字名应将名写在前而将姓写在后，比如，"杨晓峰"的英文写法为 Xiaofeng Yang。

（2）地址

写完城市之后要写中国 China。邮编的标准写法是放在省市名与国名之间，即放在 China 之前。

（3）电话

电话号码前面一定加地区号，如"86-10"；电话号码隔4位数字加一个"-"，如"6505-2126"；地区号后的括号和号码间加空格，如（86-10) 6505-2126。

2. 教育背景（Education Background）

（1）时间顺序

在英文简历中，求职者受教育情况的排列顺序是从求职者的最高教育层次写起。至于写到什么层次为止则无具体规定，可根据个人实际情况安排。

（2）学校名和地名

学校名大写并加粗，便于招聘者迅速识别你的学历。地名右对齐，全部大写并加粗，地名后一定写 China。

（3）社会工作

担任班干部只写职务就可以了，参加过社团协会应写明职务和社团名。如果什么职务都没有，写"member of club (s)"。

（4）英语和计算机水平

个人的语言水平和计算机能力应该在此单列说明。

3. 工作经历（Work Experience）

（1）时间顺序

工作经历在排列顺序上应从当前的工作岗位写起，直至求职者的第一个工作岗位为止。也有的按技能类别分类写，这主要是为了强调个人的某种技能。

（2）公司名与地名

公司名称应大写加粗。若全称太复杂，可以写得稍微简单一些，比如 International Business Machine 可简写为 IBM。

（3）公司简介

对于新公司、小公司或招聘公司不甚熟悉的某些行业的公司，略带提一下公司的简介。

（4）职务与部门

从公司名称之后的第二行开始写，职务与部门应加粗，每个词的第一个字母要大写，如 Manager，Finance Department。

（5）工作内容

1）要用点句。避免用大段文字，点句的长度以一行为宜，句数以 3~5 句为佳；点句以动词开始，目前的工作用一般现在时，以前的工作用过去时。

2）主要职责与主要成就。初级工作以及开创性不强的工作应把主要职责放在前面，而较高级或开创性较强的工作则应把主要成就写在前面。工作成就要数字化、精确化。在同一公司的业绩中，应秉持"重要优先"的原则。

3）工作时的培训。接受的培训可放在每个公司的后面，因为培训是公司内部的，与公司业务有关。

4. 兴趣爱好（Hobbies and Interests）

一般写两到三项强项就可以了，弱项一定不要写，不具体的爱好不写。

除了上述所说，写英文简历在用词上也要多加斟酌，比如有活力，用 energetic 或者 spirited 比较好，不要用 aggressive。

总之，一份好的英文简历，要目的明确、语言简练，切忌拖沓冗长，词不达意。下面是一份英文简历的示例。

Name：Yang Li

Gender：Male

Date of Birth：Feb, 1984

Citizenship：Junan county, Linyi, Shandong

Major：Architecture

➢ Education Background

September 2002 to July 2006, Henan Poly-technic Institute

September 1998 to July 2002, Linyi No. 1 Middle School

➢ Main Skills

◆ Major target

ESP of technology, e. g. architecture and real estate, and also international business.

◆ About English

Have past CET-Band 4; fluent oral English, and good pronunciation.

◆ About German

Have learned about 200 hours of Hochschuldeutsch 1, 2. Can do basic reading and writing.

◆ About Chinese

Have got the Certificate of Chinese, the score is 86.2.

◆ About Computer

Have a good command of computer, knowing the basic maintain of hardware and software.

Good sense of Visual Foxpro language. Interested in web page design, and had mastered the MACROMEDIA, which including Dreamweaver, Flash and Firework.

Skilled in Microsoft Office, including FrontPage.

➢ Work Experience

Have part time job in ShanghaiYaru Consulting Co. Ltd during 2004 and 2005 summer holiday.

Did the market research for Shandong Shiguang Boiler Co. Ltd (shanghai branch) in shanghai.

Experiences on be a tutor of junior, senior middle school students and college students.

Many times of doing promotion sales for stores during the campus life.

➢ Activities

From 2003 to 2005, be the assistant of the director of Foreign Language Department.

From 2004 to 2005, be the minister of the Network Department and Publicity Department of the Students Union.

➢ Awards & Honors

In the year of 2004, awarded the "Model student stuff" prize of our college.

In the year of 2004, won the third level scholarship of our college.

In the year of 2003, won the "Excellent Student" prize of ourcollege.

➢ Interests

Have intensive interest in traveling, photographing; Reading, esp. on business, economy and computer.

➢ Self-Evaluation

Strong sense of responsibility, good spirit of teamwork, Can learn new things well in short time.

Reading

Bluetooth Technology

Bluetooth (seen in Fig. 20-3) wireless technology has become a global technology specification for "always on" wireless communication between portable devices and desktop machines and peripherals.[1] Among the many things Bluetooth wireless technology enables users to do it swap data and synchronize files without having to cable devices together. And since the wireless link has a range of 30 feet, users have more mobility than ever before.

Bluetooth technology can also be used to make wireless data connections to conventional local area networks (LANS) via an access point equipped with a Bluetooth radio transceiver that is wired to the LAN. [2]

If desired, a user might tap out an e-mail reply on a palmtop, tell it to make an Internet connection through a mobile phone, print a copy on a printer nearby, and archive the original on the desktop PC.

The Bluetooth baseband protocol is a combination of circuit and packet switching, making it suitable for voice as well as data, Users can also connect to each other directly over a limited range through their telephones using

Fig. 20-3 Bluetooth

Bluetooth wireless technology, without incurring usage charges from a service provider.

Bluetooth wireless technology is intended to replace cable connections between computers, peripherals, and other electronic devices. In fact, one of the main advantages of Bluetooth wireless technology is that it does not need to be set up—it is always on, running in the background. [3] The Bluetooth protocols scan for other Bluetooth devices and when they can become aware of each other's capabilities, establish connections and if needed, arrange for security to protect sensitive data during transmission. The devices do not even require a line of sight to communicate with each other.

New Words and Phrases

specification　　[ˌspesɪfɪ'keɪʃn]　　n. 规格；说明书；详述
peripheral　　[pə'rɪfərəl]　　adj. 外围的；次要的；n. 外部设备，周边设备
swap　　[swɒp]　　n. 交换 vt. 与……交换；以……作交换　vi. 交换；交易
synchronize　　['sɪŋkrənaɪz]　　vt. 使……同步 vi. 同步；同时发生
mobility　　[məʊ'bɪləti]　　n. 活动性，灵活性；迁移率，机动性
conventional　　[kən'venʃ(ə)n(ə)l]　　adj. 传统的；常见的；惯例的
transceiver　　[træn'siːvə(r)]　　n. 无线电收发机，收发器
archive　　['ɑːkaɪv]　　n. 档案馆；档案文件 vt. 把……存档
protocol　　['prəʊtəkɒl]　　n. 草案，协议
transmission　　[træns'mɪʃn]　　n. 变速器；传递；传送；播送
tap out　　敲打出

Notes

[1] Bluetooth wireless technology has become a global technology specification for "always on" wireless communication between portable devices and desktop machines and peripherals.

译文：蓝牙技术已经成为便携式设备与台式机器及外围设备之间，在线无线通信的全球技术规范。

说明：此句是包含主动宾结构的简单句，只是宾语 specification 后面带有两个较长的介词短语作后置定语。

[2] Bluetooth technology can also be used to make wireless data connections to conventional local area networks (LANS) via an access point equipped with a Bluetooth radio transceiver that is wired to the LAN.

译文：蓝牙技术同样也可以用于将无线数据连接到常规局域网上，通过装有蓝牙无线收发机的接入点来进行，而此收发信机需要用线缆连接到局域网上。

说明：此句是复合句，谓语 use 在句中使用被动语态；to make 为不定式构成目的状语；via an access... transceiver 是介词短语作状语修饰 make，that 引导定语从句修饰 transceiver。

[3] In fact, one of the main advantages of Bluetooth wireless technology is that it does not need to be set up — it is always on, running in the background.

译文：事实上，蓝牙技术的主要优点是不需要进行安装，它总是连通着，在后台悄悄地运行着。

说明：此句为复合句，句中 that 引导了一个表语从句，从句中的 it 做形式主语来替代 Bluetooth wireless technology；Running in the background 是现在分词，对 always on 做进一步解释说明。

Exercises

I. Decide whether the following statements are True (T) or False (F) according to the text.

1. Bluetooth wireless technology enables users to swap data and synchronize files without having to cable devices together.

2. Bluetooth wireless link has a range of 200 feet.

3. Bluetooth technology cannot be used to make wireless data connections to conventional local area networks (LANS).

4. Bluetooth wireless technology is intended to replace cableconnections between computers, peripherals, and other electronic devices.

II. Translate the following sentences into English.

1. 蓝牙是一种无线技术标准——用来在固定和移动装置件或个人局域网之间进行短距离数据传输。

2. 蓝牙技术是具有主从结构特征的封包基础通信协议。

3. 蓝牙技术被用来取代计算机与外围设备和其他电子装置之间的有线连接。

Unit 21　Industry 4.0 Introduction

Text

Industry 4.0's provenance lies in the powerhouse of German manufacturing. [1] However the conceptual idea has since been widely adopted by other industrial nations within the European Union, and further afield in China, India and other Asian countries. The name Industry 4.0 refers to the fourth industrial revolution, with the first three coming about through mechanization, electricity and IT.

The fourth industrial revolution, and hence the 4.0, will come about via the Internet of Things and the Internet of services becoming integrated with the manufacturing environment. [2] However, all the benefits of previous revolutions in industry came about after the fact, whereas with the fourth revolution we have a chance to proactively guide the way it transforms our world.

The vision of Industry 4.0 is that in the future, industrial businesses will build global networks to connect their machinery, factories and warehousing facilities as cyber-physical systems (CPS), which will connect and control each other intelligently by sharing information that triggers actions. [3] These cyber-physical systems will take the shape of smart factories, smart machines, smart storage facilities and smart supply chains. This will bring about improvements in the industrial processes within manufacturing as a whole, through engineering, material usage, supply chains and product lifecycle management. These are what we call the horizontal value chain, and the vision is that Industry 4.0 will deeply integrate with each stage in the horizontal value chain to provide tremendous improvements in the industrial process. [4]

At the center of this vision will be the smart factory, which will alter the way production is performed, based on smart machines but also on smart products. It will not be just cyber-physical systems such as smart machinery that will be intelligent; the products being assembled will also have embedded intelligence so that they can be identified and located at all times throughout the manufacturing process. The miniaturization of RFID tags enables products to be intelligent and to know what they are, when they were manufactured, and crucially, what their current state is and the steps required to reach their desired state. [5]

This requires that smart products know their own history and the future processes required to transform them into the complete product. This knowledge of the industrial manufacturing process is embedded within products and this will allow them to provide alternative routing in the production process. For example, the smart product will be capable of instructing the conveyor belt, which production line it should follow as it is aware of it current state, and the next production process it requires to step through to completion. Later, we will look at how that works in practice.

For now, though, we need to look at another key element in the Industry 4.0 vision, and that is the integration of the vertical manufacturing processes in the value chain. The vision held is that the embedded horizontal systems are integrated with the vertical business processes (sales, logistics and finance, among others) and associated IT systems. They will enable smart factories to control the end-to-end management of the entire manufacturing process from supply chain through to services and lifecycle management. This merging of the Operational Technology (OT) with Information Technology (IT) is not without its problems, as we have seen earlier when discussing the Industrial Internet. However, in the Industry 4.0 system, these entities will act as one.

Smart factories do not relate just to huge companies, indeed they are ideal for small-sized and medium-sized enterprises because of the flexibility that they provide. For example, control over the horizontal manufacturing process and smart products enables better decision-making and dynamic process control, as in the capability and flexibility to cater to last-minute design changes or to alter production to address a customer's preference in the products design.[6] Furthermore, this dynamic process control enables small lot sizes, which are still profitable and accommodate individual custom orders. These dynamic business and engineering processes enable new ways of creating value and innovative business models.

In summary, Industry 4.0 will require the integration of CPS in manufacturing and logistics while introducing the Internet of Things and services in the manufacturing process. This will bring new ways to create value, business models, and downstream services for small medium enterprises.

New Words and Phrases

provenance ['prɒvənəns] n. 出处，起源
powerhouse ['paʊəhaʊs] n. 精力旺盛的人；发电所，动力室；强国
conceptual [kən'septjuəl] adj. 概念上的
afield [ə'fiːld] adv. 在战场上；去野外；在远处；远离
cyber ['saɪbə] adj. 计算机（网络）的，信息技术的
proactively [ˌprəʊ'æktɪvlɪ] adv. 主动地
trigger ['trɪɡə(r)] vt. 引发，引起；触发 vi. 松开扳柄 n. 扳机
assemble [ə'semb(ə)l] vt. 集合，聚集；装配；收集 vi. 集合，聚集
embed [ɪm'bed] vt. 栽种；使嵌入，使插入；使深留脑中
miniaturization [ˌmɪnətʃəraɪ'zeɪʃn] n. 小型化，微型化
tag [tæɡ] n. 标签 vt. 尾随，紧随；连接；添饰 vi. 紧随
dynamic [daɪ'næmɪk] adj. 动态的；动力学的；有活力的 n. 动态；动力
logistic [lə'dʒɪstɪkl] adj. 后勤学的
downstream [ˌdaʊn'striːm] adv. 下游地；顺流而下 adj. 下游的
accommodate [ə'kɒmədeɪt] vt. 容纳；使适应；调解 vi. 适应；调解
cater ['keɪtə(r)] vt. 投合，迎合；满足需要；提供饮食及服务
address [ə'dres] vt. 演说；写地址；向……致辞；处理 n. 地址；致辞
come about 发生；产生；改变方向
bring about 引起；使掉头
end-to-end 端对端；首尾相连
Operational Technology 经营技术；操作工艺
cyber-physical system 信息物理系统
RFID abbr. 无线射频识别 (radio frequency identification devices)
lot size 批量

Notes

[1] Industry 4.0's provenance lies in the powerhouse of German manufacturing.

译文：工业4.0起源于制造业强国德国。

说明：这是一个简单句。powerhouse 的一般含义是"精力旺盛的人或发电所"，这里取其引申意义"强国"，我们将英语翻译成中文时措辞要尽量符合汉语习惯。

[2] The fourth industrial revolution, and hence the 4.0, will come about via the Internet of Things and the Internet of services becoming integrated with the manufacturing environment.

译文：第四次工业革命，以后称之为4.0，将通过物联网和互联网服务与制造环境的集成而发生。

说明：这是一个比较复杂的简单句。The fourth industrial revolution 为主语，and hence the 4.0 为插入语，will come about 为谓语动词的将来时态，via...environment 为介词短语作方式状语来修饰 come about。在介词短语结构中，the Internet of Things and the Internet of services 是 becoming integrated with the manufacturing environment 这个动名词短语的逻辑主语。因此可称为 via + 动名词的复合结构作状语。

[3] The vision of Industry 4.0 is... by sharing information that triggers actions.

译文：工业4.0的远景是，将来的工业企业将建立全球网络来连接他们的机器、工厂和库房设施形成一个信息物理系统，这个系统将通过分享信息来引发动作，从而智能地连接各个部分并进行相互控制。

说明：这是一个复合句。The vision 为主语中心词，is 为系动词，后边的 that 引导表语从句。表语从句本身也是一个复合句。其中，industrial businesses 为主句主语，will build global networks 是动宾结构做谓语，to connect... as cyber-physical systems 是不定式结构作目的状语，而 which 引导的非限制性定语从句对其先行词 cyber-physical systems 的功能和作用起补充说明作用。

[4] These are what we call... to provide tremendous improvements in the industrial process.

译文：这是我们所称的水平价值链，对水平价值链的展望，是工业4.0将在这个价值链的每个阶段进行深度集成，从而使工业生产方法发生巨大的变化。

说明：本句为并列复合句。第一个分句是主系表结构的复合句，其表语是由 what 引导的表语从句来担当；第二个分句也是一个主系表结构的复合句，vision 在第二个分句中作主句主语，is 是系动词，that 引导表语从句。句中的 stage 是阶段的意思，指代前面提到的形成 value chain 的 engineering, material usage, supply chains, and product lifecycle management 等。to provide... 是不定式短语作目的状语。

[5] The miniaturization of RFID tags enables... to reach their desired state.

译文：缩微化的无线射频识别标签，使产品变得智能并能够知道它们是什么、什么时候被生产，更为关键的是，它们目前的状况是什么和达到期望的状态还需要的步骤。

说明：本句为复合句。The miniaturization of RFID tags 为主句的主语，enables 即主句谓语动词，products 在主句中作宾语，to be intelligent and to know... 是两个并列的不定式短语作宾语 products 的补语。而 to know 后面跟着由 what、when 和后面的 what 引导三个宾语从句作宾语。

[6] For example, control over... a customer's preference in the products design.

译文：例如，对水平制造过程和智能产品的控制，使得我们能够更好地进行决策和进行动态过程控制，在能力和灵活性方面来适应最新的设计变更或者改变生产，来满足顾客在产品设计方面的嗜好。

145

说明：这是一个复杂的简单句。control 在句子中为名词做主语中心词，后面的 over... products 为介词短语作后置定语修饰 control，enables 是谓语动词，后面跟有两个并列的宾语成分 better decision-making and dynamic process control; as in the capability and flexibility 后面的 to cater to 和 to alter 可以理解为 enable 的复合谓语动词的组成部分，to address a customer's preference in the products design 则为 production 的宾语补足语。

Exercises

Ⅰ. Answer the following questions according to the text.
1. Where is the industry 4.0 concept originated from?
2. By which technologies do the first three industry revolutions come about?
3. Via what kinds of technologies will the fourth industry come about?
4. What is the horizontal value chain of industrial process?
5. What is the vertical business processes or vertical value chain?

Ⅱ. Match the following phrases in column A with column B.

Column A	Column B
1. powerhouse	a. 标签
2. miniaturization	b. 二者选一
3. tag	c. 信息物理系统
4. alternative	d. 缩微化
5. mechanization	e. 物联网
6. Internet of Things	f. 发电所；强者
7. end-to-end	g. 机械化
8. cyber-physical system	h. 首尾相连

Ⅲ. Translate the following sentences into Chinese.

1. The first industrial revolution mobilized the mechanization of production using water and steam power.

2. The Second Industrial Revolution dates from Henry Ford's introduction of the assembly line in 1913, which resulted in a huge increase in production.

3. The Third Industrial Revolution resulted from the introduction of the computer onto the factory floor in the 1970s, giving rise to the automated assembly line.

4. The vision of Industry 4.0 is for "cyber-physical production systems" in which sensor-laden "smart products" tell machines how they should be processed.

Practical English

怎样写英文求职信

1. 信函的内容

1) 写信的目的或动机：求职信上一定要说明写信的缘由和目的。

2) 个人资料：写信人应说明自己的年龄或出生年月、教育背景，尤其是和应征职位有关的训练或教育科目、工作经验或特殊的技能。如无实际经验，略述在学类似经验亦可。

3) 结尾：求职信的结尾可以希望未来的雇主给予面谈的机会，因此，信中要说明可以面谈的时间。

2. 机智和良好的判断

写求职信要有机智和良好的判断，下列几点可供参考。

1）陈述事实，避免表示意见。

2）不要批评他人。

3）不要过分渲染自我。

3. 英文求职信的表达

1）应征函的第一段说明写信的缘由和目的，不宜用分词从句，下面的句子较为恰当。

In your advertisement for an accountant, you indicated that you require the services of a competent person, with thorough training inthe field of cost accounting. Please consider me an applicant for the position. Here are my reasons for believing I am qualified for this work.

2）提到待遇时，不要过分谦虚或表示歉意。

I feel it is presumptuous of me to state what my salary should be. My first consideration is to satisfy you completely. However, while I am serving my apprenticeship, I should consider ￥2,600 a month satisfactory compensation. 我不敢冒昧说出起薪多少。最初我仅想要如何工作得好，使您满意。在试用期间，月薪两千六百元即可。

3）结尾部分应尽量避免陈腐、太过自信或太过具体。

4. 求职信的语气

求职信要发挥最大的效果，语气必须肯定、自信、有创意而不过分夸张。可以比较下面的句子：

I am confident that my experience and references will show you that I can fulfill the particular requirements of your bookkeeping position. 我相信我的经验和推荐人可以告诉您，我能够符合贵公司簿记员一职的特定需要。（有特殊风格）

I feel quite certain that as a result of the course in filing which I completed at the Crosby School of Business, I can install and operate efficiently a filing system for your organization. 我相信在克洛斯比商业实习班修完一门课后，我能够替贵公司设置并且操作一套档案分类系统。（比较谦虚）

要写好英文求职信，除了上述所说各点之外，还要注意纸张的选用和格式排列，语法、标点和拼写要正确无误。这些要求跟写中文求职信是一样的。下面是一份英文求职信的示例。

April 6, 2010

P. O. Box 3

Henan Poly-technic Institute

Nanyang, China 473009

Dear Sir/Madame:

Your advertisement for a Network MaintenanceTechnician in the April 10 Student Daily interested me, because the position that you described sounds exactly like the kind of job I am seeking.

According to the advertisement, your position requires a good college student, Bachelor or above in Computer Science or equivalent field and proficient in Windows NT 4.0 and Linux System. I feel that I am competent to meet the requirements. I will be graduating from Henan Poly-technic Institute this year with my B. S. degree and BEC Vantage. My studies have included courses in computer control and management and I designed a control simulation system developed with Microsoft Visual and SQL Server.

During my education, I have grasped the principles of my major subject area and gained practical skills. Not only have I passed CET-Band 4 and BEC Vantage, but more importantly I can communicate fluently in English. My ability to write and speak English is a good standard.

I would welcome an opportunity to attend you for an interview.

Enclosed is my resume and if there is any additional information you require, please contact me (Tel: ××××××××).

Yours faithfully

×××

Reading

Made in China 2025 and Industrie 4.0 Cooperative Opportunities

Following the Chinese government's issuance of its *Made in China 2025* strategy, which outlines plans to upgrade the mainland's industries, its 13th Five-Year Plan, adopted in March 2016, sets out to deepen the implementation of this strategy in the next five-year period (2016-2020).[1] While this has aroused interests as regards the development direction of Chinese industry, some industry observers have drawn parallels with Germany's Industrie 4.0 strategy, which was designed to enhance the efficiency of German industry.

It is worth noting that some have raised concerns that the two strategies may lead to intensified competition between Chinese and German industries.[2] Nevertheless, the two countries signed a memorandum of understanding to step up cooperation in the development of smart manufacturing technology in July 2015.

Indeed, the relative industrial development of the two countries, coupled with different strategic development priorities, reveal more opportunities for cooperation than competition, including in the area of industrial robots. Moreover, the different positions held by Chinese and German industries in the global supply chain also hint at further opportunities for relevant players stemming from Sino-foreign cooperation projects.

Essentially, Germany's Industrie 4.0 advocates the adoption of state-of-the-art information and communication technology in production methods as a means to further enhance industrial efficiency. This strategy is developed on the basis that Germany's strong machinery and plant manufacturing industry, its IT competences and expertise in embedded systems and automation engineering make it well placed to consolidate its position as a global leader in the manufacturing engineering industry.

Industrie 4.0 aims for intelligent production by connecting the current embedded IT system production technologies with smart processes in order to transform and upgrade industry value chains and business models. This will require Germany to enhance its research and development efforts in areas such as further integrating manufacturing systems. New industry and technical standards will be required to enable connections between the systems of different companies and devices, while data security systems will need to be upgraded to protect information and data contained in the system against misuse and unauthorized access. All of these developments are expected to enhance the efficiency and innovative capacity of German industry, while saving resources and costs.

As regards *Made in China 2025*, the focus is on innovation and quality, as well as guiding Chinese industries to move away from low value-added activities to medium- and high-end manufacturing operations, rather than pursuing expansion of production capacity. The strategy is also aimed at eliminating inefficient and outdated production capacity, and helping enterprises to conduct more own-design and own-brand business. These objectives are to be facilitated by actions including the establishment of manufacturing innovation centers, strengthening intellectual property rights protection, building up new industrial standards, and facilitating the development of priority and strategic sectors.

New Words and Phrases

Industrie　　［ˈɪndəstɪ］　　n. 工业（德语）
cooperative　　［kəʊˈɒpərətɪv］　　adj. 合作的；协助的；共同的
issuance　　［ˈɪʃjuːəns］　　n. 发布，发行
arouse　　［əˈraʊz］　　vt. 引起；唤醒；鼓励　vi. 激发；醒来
memorandum　　［meməˈrændəm］　　n. 备忘录；便笺
tap　　［tæp］　　vt. 轻敲；轻打　vi. 轻拍　n. 水龙头；轻打　vt. 采用；开发
chancellor　　［ˈtʃɑːnsələ(r)］　　n. 总理（德、奥等的）；（英）大臣；校长
stem　　［stem］　　n. 干；茎；血统　vt. 阻止　vi. 阻止；起源于某事物；逆行
advocate　　［ˈædvəkeɪt］　　vt. 提倡，主张，拥护　n. 提倡者；支持者；律师
consolidate　　［kənˈsɒlɪdeɪt］　　vt. 巩固，使固定；联合　vi. 巩固，加强
facilitate　　［fəˈsɪlɪteɪt］　　vt. 促进；帮助；使容易
intellectual　　［ˌɪntəˈlektʃʊəl］　　adj. 智力的；聪明的；理智的　n. 知识分子
in line with　　符合；与……一致
hint at　　暗示；对别人暗示……
state-of-the-art　　最先进的；已经发展的；达到最高水准的

Notes

［1］Following the Chinese government's issuance of its *Made in China 2025* strategy, which outlines plans to upgrade the mainland's industries, its 13th Five-Year Plan, adopted in March 2016, sets out to deepen the implementation of this strategy in the next five-year period (2016-2020).

译文：随着中国政府对升级大陆工业的纲要规划《中国制造2025》战略的发布，2016年3月采用的"十三五"规划已经开始该战略在下一个五年计划（2016—2020）的纵深部署。

说明：这是一个复合句。主句的主语是 its 13th Five-Year Plan；sets out 是主句谓语动词；后面的不定式短语 to deepen... 作宾语；Following... 是现在分词短语作时间状语；其中 which 引导非限制性定语从句，对 strategy 起补充说明作用；adopted in March 2016 是过去分词短语作 13th Five-Year Plan 的后置定语。

［2］It is worth noting that some have raised concerns that the two strategies may lead to intensified competition between Chinese and German industries.

译文：值得一提的是已经有人提出了担心，这两个战略将使得中国和德国工业之间的竞争加剧。

说明：这是一个含有主语从句的复合句。句中第一个 that 引导主语从句，some have

raised concerns 作后置主语，it 为形式主语；句中第二个 that 引导同位语从句，对 concerns 起解释说明作用。

Exercises

Ⅰ. Decide whether the following statements are True (T) or False (F) according to the text.

1. There is no cooperation between China and German governments in updating their industry because of concerns that the two strategies may lead to intensified competition between Chinese and German industries.

2. The relative industrial development of China and Germany, coupled with different strategic development priorities, render more opportunities for cooperation than competition.

3. Germany has strong competences and expertise in embedded systems and automation engineering.

4. For better communication and connecting between different systems, some new industrial and technical standards need to be developed.

5. The priorities of Germany's Industrie 4.0 and China's Made in China 2025 strategies are different.

Ⅱ. Translate the following sentences into English.

1. 工业 4.0 还没有成为现实，它还只是一个理念，但它可能将带来深远的变化和影响。

2. 预测所有的环境变化是不可能的，而这些变化又是控制系统动态响应所必需的，所以可编程逻辑将变得非常重要。

3. 无论是革命性的变化还是不断演变，工业化生产将变得更加高效。

Unit 22 3D Printing

Text

Additive Manufacturing (AM) is an appropriate name to describe the technologies that build 3D objects by adding layer-upon-layer of material, whether the material is plastic, metal, concrete or one day... human tissue. [1]

Common to AM technologies is the use of a computer, 3D modelling software (Computer Aided Design or CAD), machine equipment and layering material. Once a CAD sketch is produced, the AM equipment reads in data from the CAD file and lays downs or adds successive layers of liquid, powder, sheet material or other, in a layer-upon-layer fashion to fabricate a 3D object.

The term AM encompasses many technologies including subsets like 3D Printing, Rapid Prototyping (RP), Direct Digital Manufacturing (DDM), layered manufacturing and additive fabrication.

AM application is limitless. Early use of AM in the form of Rapid Prototyping focused on pre-production visualization models. More recently, AM is being used to fabricate end-use products in aircraft, dental restorations, medical implants, automobiles, and even fashion products.

While the adding of layer-upon-layer approach is simple, there are many applications of AM technology with degrees of sophistication to meet diverse needs including: a visualization tool in design, a means to create highly customized products for consumers and professionals alike, as industrial tooling, to produce small lots of production parts, one day... production of human organs.

At MIT, where the technology was invented, projects abound supporting a range of forward-thinking applications from multi-structure concrete to machines that can build machines;[2] while work at Contour Crafting supports structures for people to live and work in.

Some envision AM as a compliment to foundational subtractive manufacturing (removing material like drilling out material) and to lesser degree forming (like forging). Regardless, AM may offer consumers and professionals alike, the accessibility to create, customize and/or repair product, and in the process, redefine current production technology.

Whether simple or sophisticated, AM is indeed amazing and best described in the adding of layer-upon-layer, whether in plastic, metal, concrete or one day... human tissue. [3]

Here are some Examples of Additive Manufacturing.

SLA (Stereo lithography Appearance)

This is a very high end technology utilizing laser technology to cure layer-upon-layer of photopolymer resin (polymer that changes properties when exposed to light).

The build occurs in a pool of resin. A laser beam, directed into the pool of resin, traces the cross-section pattern of the model for that particular layer and cures it. During the build cycle, the platform on which the build is repositioned, lowers by a single layer thickness. The process repeats until the build or model is completed and fascinating to watch. Specialized material may be needed to add support to some model features. Models can be machined and used as patterns for injection molding, thermoforming or other casting processes.

FDM (Fused Deposition Modelling)

Process oriented involving use of thermoplastic (polymer that changes to a liquid upon the application of heat and solidifies to a solid when cooled) materials injected through indexing nozzles onto a platform.[4] The nozzles trace the cross-section pattern for each particular layer with the thermoplastic material hardening prior to the application of the next layer. The process repeats until the build or model is completed and fascinating to watch. Specialized material may be needed to add support to some model features. Similar to SLA, the models can be machined or used as patterns.

MJM (Multi-Jet Modelling)

Multi-Jet Modelling is similar to an inkjet printer in that a head, capable of shuttling back and forth (3 dimensions—x, y, z), incorporates hundreds of small jets to apply a layer of thermopolymer material, layer-by-layer.

3DP (Three-Dimensional Printing)

This involves building a model in a container filled with powder of either starch or plaster based material. An inkjet printer head shuttles and applies a small amount of binder to form a layer. Upon application of the binder, a new layer of powder is swept over the prior layer with the application of more binder.[5] The process repeats until the model is complete. As the model is supported by loose powder there is no need for support. Additionally, this is the only process that builds in colours.

SLS (Selective Laser Sintering)

Somewhat like SLA technology Selective Laser Sintering (SLS) utilizes a high powered laser to fuse small particles of plastic, metal, ceramic or glass. During the build cycle, the platform on which the build is repositioned, lowers by a single layer thickness. The process repeats until the build or model is completed. Unlike SLA technology, support material is not needed as the build is supported by unsintered material.

New Words and Phrases

additive ['ædɪtɪv] adj. 附加的；[数] 加法的 n. 添加剂，添加物
concrete ['kɒŋkriːt] adj. 混凝土的；实在的，具体的 n. 混凝土
tissue ['tɪʃuː; 'tɪsjuː] n. 组织；纸巾
sketch [sketʃ] n. 素描；略图；梗概
successive [sək'sesɪv] adj. 连续的；继承的；依次的；接替的
powder ['paʊdə] n. 粉；粉末 vt. 使成粉末；撒粉 vi. 搽粉；变成粉末
fabricate ['fæbrɪkeɪt] vt. 制造；伪造；装配
encompass [ɪn'kʌmpəs; en-] vt. 包含；包围，环绕；完成
abound [ə'baʊnd] vi. 富于；充满
subset ['sʌbset] n. 子集
prototype ['prəʊtətaɪp] n. 原型；标准，模范
visualization [ˌvɪzjʊəlaɪ'zeɪʃən] n. 形象化；清楚地呈现在心
dental ['dent(ə)l] adj. 牙科的；牙齿的，牙的
implant [ɪm'plɑːnt] vt. 种植；嵌入 vi. 被移植 n. [医] 植入物；植入管
contour ['kɒntʊə] n. 轮廓；等高线；周线；概要 vt. 画轮廓；画等高线
envision [ɪn'vɪʒn] vt. 想象；预想
subtractive [səb'træktɪv] adj. 减去的；负的；有负号的

customize   ['kʌstəmaɪz]   vt. 定做，按客户具体要求制造
photopolymer   [ˌfəʊtəʊ'pɒlɪmə]   n. 光聚合物，光敏聚合物；感光性树脂
polymers   ['pɒlɪməs]   n. [高分子]聚合物
pattern   ['pæt(ə)n]   n. 模式；图案；样品
thermoforming   [θəmə'fɔːmɪŋ]   n. 热成型；热压成形
cast   [kɑːst]   vt. 浇铸；投，抛；计算；投射（光、影、视线等）
orient   ['ɔːrɪənt; 'ɒr-]   vt. 使适应；确定方向   n. 东方   adj. 东方的
thermoplastic   [θɜːməʊ'plæstɪk]   adj. 热塑性的   n. [塑料]热塑性塑料
indexing   ['ɪndeksɪŋ]   n. 指数化；【机械学】分度，转位
nozzle   ['nɒz(ə)l]   n. 喷嘴；管口；鼻
deposition   [ˌdepə'zɪʃ(ə)n; diː-]   n. 沉积物；矿床；革职
shuttle   ['ʃʌtl]   n. 梭子；航天飞机   vt. & vi. 穿梭般来回移动
incorporate   [ɪn'kɔːpəreɪt]   vt. 组成公司；包含   vi. 包含；吸收；合并
starch   [stɑːtʃ]   n. 淀粉
plaster   ['plɑːstə(r)]   n. 灰泥，涂墙泥；石膏；膏药
binder   ['baɪndə(r)]   n. 黏结剂；包扎物，包扎工具；装订工
selective   [sɪ'lektɪv]   adj. 精心选择的；选择的，不普遍的；淘汰的
sintering   ['sɪntərɪŋ]   n. 烧结   v. 烧结；使熔结
ceramic   [sɪ'ræmɪk]   adj. 陶瓷的；陶器的；n. 陶瓷；陶瓷制品

Technical Terms

Additive Manufacturing   增材制造
Rapid Prototyping   快速造型
Rapid fabrication   快速制造
Direct Digital Manufacturing   直接数字制造
Selective Laser Sintering   选择性激光烧结
Fused Deposition Modelling   熔融沉积造模
Multi-Jet Modelling   多重喷射造模
Contour Crafting   轮廓工艺
Three-Dimensional Printing   3D 打印
cross-section   剖面

Notes

[1] Additive Manufacturing (AM) is an appropriate name to describe the technologies that build 3D objects by adding layer-upon-layer of material, whether the material is plastic, metal, concrete or one day... human tissue.

译文：增材制造（AM）是描述这种 3D 物体生成技术的一个恰当名字：通过一层一层地添加材料来生成物体，这些材料可以是塑料、金属、水泥，甚至有一天会使用人体的组织。

说明：这是一个复合句。句中 that build 3D... material 是定语从句，修饰 technologies；whether 则引导状语从句修饰 build。

[2] At MIT, where the technology... concrete to machines that can build machines.

译文：在 AM 技术的发明地麻省理工学院，有大量的项目来支持一系列的前瞻性的应用，从多结构混凝土到可以制造机器的机器。

说明：这是一个复合句。主句主语是 projects；abound 是谓语动词，"富于，充满"的意思；At MIT 为地点状语，where 引导的定语从句修饰 MIT；supporting a range of … machines 为分词短语做状语，修饰动词 abound；该分词短语中 that 引导的定语从句修饰它前面的 machines。

[3] Whether simple or sophisticated, AM is indeed amazing and best described in the adding of layer-upon-layer, whether in plastic, metal, concrete or one day… human tissue.

译文：无论是简单还是复杂，AM 确实令人惊叹，以一层一层添加的方式对这个令人惊叹的工艺进行了最好呈现，无论是采用塑料、金属、混凝土或者有一天……人体组织。

说明：这是一个包含有两个让步状语的简单句。AM 是主语，is 为系动词，amazing and best described 分别为形容词和动词过去分词做表语，whether (the material adopted is) in plastic, metal, concrete or one day… human tissue 为让步状语，可以看作在 whether 后边省略了 the material adopted is；而在句首 Whether simple or sophisticated，可以看成在 Whether 后面省略了（it is）的让步状语，用来修饰主句谓语 is indeed amazing。

[4] Process oriented involving use… through indexing nozzles onto a platform.

译文：这种工艺定位于使用热塑性塑料（当施加热量时变成液体并在冷却时固化成固体的聚合物），通过转位喷嘴注射到平台上。

说明：去掉括号内容，可以看成这是一个包含主动宾结构的简单句。Process 是主语，oriented 是谓语动词，宾语是 involving use… materials，之后的 injected through… 是过去分词短语作定语，修饰 materials。括号中由 that 引导的定语从句修饰 polymer，而 polymer 是 thermoplastic 的同位语，都是对 thermoplastic 起解释说明作用的。注意 indexing 在句中的用法，index 在机械技术英语中一般是"转位的，分度的"意思。比如：indexing table，分度盘；indexable insert，可转位刀片。

[5] Upon application of the binder, a new layer… prior layer with the application of more binder.

译文：当一层黏结剂施加完成，马上在这一层黏结剂上撒一层粉末，然后施加另一层黏结剂。

说明：这是一个比较复杂的简单句。Upon application of the binder 在句首是时间状语；upon 的常见意思是"当……时候"，或者"一……就……"，意思与 when 相当。但是注意，when 是副词，而 upon 是介词，后面只能跟名词结构。

Exercises

I. Answer the following questions according to the text.

1. What is Additive Manufacture?
2. What is common to AM technologies?
3. What is the focus of early use of AM technology in the form of Rapid Prototyping?
4. Where did the AM technology invented?
5. What kind of material can be used in the process of SLA?

II. Match the following phrases in column A with column B.

Column A   Column B
1. polymer   a. 剖面

2. laser
3. cross-section
4. Direct Digital Manufacturing
5. Additive Manufacturing
6. Selective Laser Sintering
7. Rapid Prototyping
8. Rapid Fabrication

b. 直接数字制造
c. 增材制造
d. 选择性激光烧结
e. 激光
f. 高分子
g. 快速制造
h. 快速造型

Ⅲ. Translate the following sentences into Chinese.

1. Everybody recognizes the term 3D printing nowadays but what many people mean when talking about 3D printing is actually one of the several Additive Manufacturing (AM) processes.

2. Individual Additive Manufacturing processes differ depending on the material and machine technology used.

3. You will have to prepare a 3D digital model before it is ready to be 3D printed.

4. The term "3D printing" 's origin sense is in reference to a process that deposits a binder material onto a powder bed with inkjet printer heads layer by layer.

Practical English

面 试 技 巧

如何在英语面试中脱颖而出？在面试中应该避免哪些内容？下面总结了英语面试中的几点技巧和具体应对策略，希望能对面试者有所帮助。

1. 面试技巧

（1）面试中要自信

英语面试以考查英语应用能力为主，小组讨论是英语面试的重要一环。一般来说，小组讨论形式的面试目的主要是要了解应聘者在团体中的工作反应，所以应聘者在面试中最需要的是自信、敢说。注意在互动过程中要多使用口语化的表达，尽量采用简短明了的词汇来表达自己的意思，说话要流利，思维要连贯，层次要清晰，可以用"well""however"这样的过渡词来给自己停顿和思考。

（2）面试前了解一下企业情况

在面试前不妨去公司的英文网站了解一下企业情况。必须准确了解应聘公司的外语使用程度和目的等信息，并做出相应的对策。

（3）注意英语时态的变化运用

面试过程中，往往会涉及很多关于个人经历、教育背景、工作经验和职业规划等方面的问题，因此在表述某件事情或是某个想法的时候，一定要注意配合正确的时态，否则就会差之毫厘，失之千里。例如，你已经参加过某项专业技能培训与你正在参加或计划参加就是完全不同的。

（4）尊重个人及文化差异

主要有两种做法要特别注意避免。首先是要避免使用过于生僻的单词，或者使用地方俚语之类接受群体相对比较小的表达方式，因为这种表达方式很有可能造成听者的困惑与曲解。其次则是要避免过多、过于主观地谈及宗教文化或时事政治方面的问题，因为面试官很有可能来自不同的国家与地区，有一定的个人倾向。因此作为面试者，在不了解情况的状态下，如果谈到此类话题，要谨慎而有节制的发言。

2. 具体应对策略

（1）介绍自己

可以从以下几个方面突出自己：职业技能强项、该技能的深入程度、使用该技能完成过的项目以及其他相关方面的深入研究和学习探索，注意突出个性。不要在自己的姓名、年龄、毕业学校、兴趣爱好等方面浪费时间，也不用再把你的简历背诵一遍。

（2）应聘动机

在回答这个问题的时候要充满信心，不要说是来试试的（Do you really want to take a try?）。要事先将公司信息整理好，和自己的职业技能相结合来介绍。

（3）教育背景

回答这个问题要简明扼要，实话实说。

（4）工作经验

回答此问题要充分显露自己的实力，但同时要避免给人炫耀之感。

（5）工作要求

要充分表现能胜任的信心。

（6）性格爱好

要表现积极合作的一面。

（7）弱点

客观地评价一下自己确实存在但对工作并不重要的缺点是一个不错的办法，但要提一下将采取什么样的办法来克服或弥补自己的缺点。

（8）过去的工作

面试官想通过你对过去的工作的看法，了解你对今后的工作态度。不要说你不喜欢过去的工作，可以谈谈过去的工作特点和你所负责的主要任务，但要突出过去的工作趋于平淡且失去挑战性，你可以在更艰巨的任务和岗位中发挥能力。

（9）将来的打算

可以用简洁的几句话表达希望在这个公司一直工作下去，同时要体现出上进心，渴望与公司一起发展和进步，不断提升自己。

An Interview for an Export Trader（招聘外贸员面试范例）
（I：Interviewer；A：Applicant）

A：Good afternoon, sir.

I：Good afternoon. Please take a seat.

A：Thank you.

I：You areFeida Ning? I am Henry Hudson.

A：Yes. Nice to see you, Mr. Hudson.

I：To start with, tell me about your education, please.

A：All right. I graduated from the foreign language department in Henan Polytechnic Institute three years ago. I majored in international trade.

I：Very well. I see from your resume that you have been working for an import and export company in Hangzhou since your graduation from college. What is your chief responsibility there?

A：I am responsible for exporting light industrial machinery to some Asian and European countries.

I：Have you traveled a lot in your work?

A: Yes. I have traveled dozens of times abroad. I have been to such countries as Thailand, Singapore, Japan, Indonesia, Burma, the Netherlands, Denmark, Italy, Germany and England to do business.

I: Are you single or married?

A: I'm still single. Nowadays many young people in China are not in a hurry to get married. They'd rather secure their careers before they settle down in a family.

I: That's the kind of man we are looking for. Our promotion work entails much travel. So we need employees without family burdens yet. Now tell me if you have a good command of both written and spoken English.

A: When I was at college, I passed Band Six of College English Test. I also passed Advanced Business English Certificate Test. All the foreign businessmen I've dealt with say my English is quite good.

I: May I ask why you want to change jobs?

A: Because I wish to get a more challenging opportunity at a foreign capital company.

I: Why are you interested in this company?

A: A friend of mine works here, and he told me about your company, so I became interested. I think working in this company would provide me with a good opportunity to use my knowledge.

I: What do you know about this company?

A: This company is one of the biggest manufacturing companies in the world. There're a lot of branches in all parts of the world with the head office in the USA IBM (China) Co. Ltd. was set up in Beijing in 1992. It has established branches in Shanghai, Guangzhou, Hangzhou, Shenyang, Shenzhen, Nanjing, Wuhan and Xi'an. It deals in business machines.

I: Do you know what GMFNT stands for?

A: Of course. It stands for General Most-Favored-Nation Treatment. If one nation enjoys this kind of treatment, it is accessible to tariff preference for imported goods from another nation.

I: Now I'm going to ask you a few professional questions. What is the first thing to do in international trade?

A: As a buyer, you first have to make an inquiry. And as a seller, you first have to make a marketing research according to the cost audit.

I: Can you name some terms of payment?

A: Of course. Irrevocable letter of credit, confirmed letter of credit, and transferable and divisible letter of credit are common terms of payment in international trade.

I: You are right. We'll notify you of our final decision within one week.

A: Thank you, Mr. Hudson, for your interview with me. You can Email your decision to me. I hope to see you again.

Reading

American Scientists Work on Printing of Living Tissue Replacements

Three-dimensional printers work very much like ordinary desktop printers. But instead of just putting down ink on paper, they stack up layers of living material to make 3-D shapes. The technology has been around for almost two decades, providing a shortcut for dentists, jewelers, machinists and even chocolate producer who want to make custom pieces without having to create molds. [1]

In the early 2000s, scientists and doctors saw the potential to use this technology to construct living tissue, maybe even human organs. They called it 3-D bioprinting, and it is a red-hot branch of the burgeoning field of tissue engineering.

In laboratories all over the world, experts in chemistry, biology, medicine and engineering are working on many paths toward an audacious goal: to print a functioning human liver, kidney or heart using a patient's own cells. That's right—new organs, to go. If they succeed, donor waiting lists could become a thing of the past.

Bioprinting technology is years and possibly decades from producing such complex organs, but scientists have already printed skin and vertebral disks and put them into bodies. So far, none of those bodies have been human, but a few types of printed replacement parts could be ready for human trials in two to five years.

Scientists say the biggest technical challenge is not making the organ itself, but replicating its intricate internal network of blood vessels, which nourishes it and provides it with oxygen. [2]

Many tissue engineers believe the best bet for now may be printing only an organ's largest connector vessels and giving those vessels' cells time, space and the ideal environment in which to build the rest themselves; after that, the organ could be implanted.

New Words and Phrases

stack [stæk] vt. & vi. 堆成堆，垛；堆起来或覆盖住
jeweler [ˈdʒuːələ] n. 珠宝商；宝石匠
machinist [məˈʃiːnɪst] n. 机械师
red-hot [ˈredˈhɔt] adj. 赤热的；激烈的，恼怒的；近期的，新的
burgeoning [ˈbɜːdʒənɪŋ] adj. 迅速成长的；v. 迅速发展；发（芽）
audacious [ɔːˈdeɪʃəs] adj. 大胆的；鲁莽的；大胆创新的
functioning [ˈfʌŋkʃənɪŋ] v. 起作用（function 的现在分词）；正常工作
liver [ˈlɪvə(r)] n. 肝脏
kidney [ˈkɪdni] n. 肾，肾脏
donor [ˈdəʊnə(r)] n. 捐赠者
vertebral [ˈvɜːtɪbrəl] adj. 椎骨的；脊椎的
replicate [ˈreplɪkeɪt] vt. 复制，复写；[生] 复制
intricate [ˈɪntrɪkət] adj. 错综复杂的；难理解的；曲折；盘错
nourish [ˈnʌrɪʃ] vt. 滋养，施肥于；抚养，教养；使健壮
vessel [ˈvesəl] n. 血管，脉管；容器，器皿

Notes

[1] The technology has been... to make custom pieces without having to create molds.

译文：这项技术已经有将近 20 年的历史，为牙医、珠宝商以及机械师提供了更为快捷的手段，甚至还为那些希望不用模具就可以为人们定制巧克力的生产商实现了梦想。

说明：这是一个复合句。主句的主语是 The technology，has been 为系动词的完成形式，around for almost two decades 是表语部分；Providing shortcut... to create molds 是现在分词短语作伴随情况的状语，其中 who 引导定语从句修饰 producer。

[2] Scientists say the biggest technical challenge... and provides it with oxygen.

译文：科学家称最大的技术挑战并不是制造器官本身，而是复制器官内部错综复杂的血管网络，这些血管起到了滋养器官、为器官提供氧气的作用。

说明：这是包含有宾语从句的复合句。句中 Scientists 是主语，say 是谓语动词，the biggest technical ... with oxygen 前面省略了 that，是宾语从句作宾语；在宾语从句中使用了 is not ... but 表示"不是……而是……"；which 引导的非限制性定语从句修饰 blood vessels，对先行词起补充说明作用。

Exercises

Ⅰ. Decide whether the following statements are True (T) or False (F) according to the text.

1. 3D printing is a technology which was invented five years ago.
2. Scientists and doctors saw the potential to use this technology to print human tissues 10 years ago since the technology developed not very fast.
3. There has already been a successful case to print human organs by 3D printers.
4. The biggest technical challenge to print a human organ is to replicate its intricate internal network of blood vessels.

Ⅱ. Translate the following sentences into English.

1. 3D 打印机从 20 世纪 80 年代就已经存在并被广泛应用于快速造型和研发用途。
2. 3D 打印已经被考虑作为能够生成人类活体新组织和器官的干细胞的移植方法。
3. STL 文件从 20 世纪 80 年代中期起就是增材设备和设计程序间传递信息的标准。

Appendix

Appendix A   Communication Skills Training for Careers
（职场交际技能训练）

Dialogue 1   Entering the Company

Robert: Hello, sir. I'm a newcomer in our company. Glad to meet you here.
James: Hello. Nice to meet you.
Robert: Are you Mr. James?
James: Yes, I am. What about you?
Robert: I'm Gill Robert. I have heard a lot about you.
James: Really? Nothing bad, I hope.
Robert: Of course not. You made a lot of new equipment by yourself.
James: Yes. I like this work.
Robert: OK! I will talk to you later.

Related Expressions

1. Hi! How are you? / How are you doing? 你好吗？
2. Hi! Nice to meet you! 你好，很高兴见到你。
3. Fancy meeting you here. 真想不到在这儿碰上你。
4. I don't think I've had the pleasure. 我想我未曾有幸见过你。
5. Hello, Mr. Wu, let me introduce you to Mr. Liu. He is our new manager of our Marketing Department. 你好，吴先生，让我给你介绍一下刘先生，他是我们营销部的新经理。
6. I've heard a lot about you, Betty, now I finally have the chance to meet you in person. 贝蒂，久仰大名，今天终于有幸见到你了。

Dialogue 2   Making an Appointment

Secretary: Good morning, Odyssey Promotions. May I help you?
Student: This is Li Ming. I'd like to speak to Bruce Baker, please. [1]
Secretary: I am sorry. Mr. Baker is not available now. Would you like to leave a message for him? [2]
Student: Well, I'd like to make an appointment with him for a job interview these days.
Secretary: Then, do you have any particular time in mind?
Student: How about this Friday?
Secretary: I'm afraid that Mr. Baker will not be available on that day.
Student: In that case, what about next Monday?
Secretary: One moment, please let me check Mr. Baker's schedule. Yes, he will be available

next Tuesday morning. Would nine o'clock be all right?

Student: OK. I'll visit him at 9:00 next Tuesday.

Secretary: That would be fine.

Student: Thank you. Goodbye.

Secretary: Bye.

Notes

[1] I'd like to speak to Bruce Baker, please. 译为"我想和布鲁斯·贝克通话,可以吗?"。类似的还有: Could I speak to Mr. Parley? (我能和帕利先生通话吗?) Have I the pleasure of addressing Mr. Chris? (我可以和科里斯先生讲话吗?)

[2] Would you like to leave a message for him? = Can I take a message? 您想留个口信吗?

Dialogue 3 Amplifier Design

Supervisor: I would like you to do a preliminary calculation for an amplifier.

Engineer: What are the details? How is the amplifier to be used?

Supervisor: It must give a voltage gain of about 1000.

Engineer: That isn't enough information. I need to know the desired input impedance as well as the types of loads to be used with the amplifier.

Supervisor: It must have ahigh input impedance with an input capacitance of no more than 50 microfarads. The loads will be resistive in the frequency to be used. A few kilo-ohms will be enough, I believe.

Engineer: What will be the frequency?

Supervisor: About one megacycle with a 200-kilocycle bandwidth.

Engineer: How large will the input signals be?

Supervisor: About half a volt, I believe. But smaller signals may be possible.

Engineer: Any weight or size restrictions? Can we use transistors, or would you preferICs?

Supervisor: For the time being, there are no restrictions.

Engineer: Is there any particular time when I must finish it?

Supervisor: I'd like to test the first model in about a week.

Engineer: Ok, I will begin the work right now.

Related Expressions

1. How soon can you have your product ready? 你们多久可以把产品准备好?

2. When can you effect shipment? I'm terribly worried about late shipment. 你们什么时候交货? 我非常担心货物迟交。

3. Here is my business card. Please contact me if you are interested in our products. 这是我的名片,如果您对我们的产品感兴趣,请与我联系。

4. Our products are quite competitive in quality and price. 我们的产品优质价廉。

5. It must have ahigh input impedance with an input capacitance of no more than 50 microfarads. 它必须拥有不超过50pF输入电容的高输入电阻。

Dialogue 4 Repairing a TV Set

Clark: What's the matter with your TV set?

Brown: It failed to work. I hope it's nothing serious.

Clark: I don't think so. First I'll unscrew the back plate.

Brown: Look at all those resistors, capacitors, transistors and wires, it's pretty complicated.

Clark: Yes.

Brown: Can I plug it in now?

Clark: Not yet.

Brown: How are you going to find out what's wrong?

Mark: Give me the multimeter.

Brown: OK.

Clark: I see the problem. It's loose wire. Hand me the soldering gun and some solder.

Brown: Here you are.

Clark: Thanks. That should do it. Turn it on.

Brown: Great! It can play again. Thank a lot.

Clark: Don't mention it.

Related Expressions

1. The machine seems to be out of gear. 机器似乎出故障了。

2. These instructions should be followed strictly. 应严格遵循操作说明。

3. These instruments differ in their precision after operation. 经过操作，这些仪器的精确度大不相同。

4. Please have the connections checked, and see if there are any wrong connections. 请检查线路，看是否有线搭错。

5. Tighten the screws properly. 拧紧螺钉。

6. First I'll unscrew the back plate. 首先我要打开电视的后盖。

7. Hand me the soldering gun and some solder. 请递给我焊枪和一些锡线。

Dialogue 5　Talking About the Internet

Dick: What is the computer network like? Is it a bunch of computers hooked together?

Jack: You may say so, but to be more exact, it's a collection of computers, printers, and other devices that can communicate directly with each other.

Dick: Then how are they connected, through telephone wires?

Jack: Well, mostcomputer networks are local area networks and they are connected by special cables or phone wires.

Dick: I see. No wonder they say we can dial up the network by way of modern connection.

Jack: Sure, But that's only part of its power. More important is that you can also get from other computers or share information between computers on the network.

Dick: Talking of information sharing on the network, does it mean that I can see what's on someone else's hard drive in another place?

Jack: That's true, and youcan share his or her CD-ROM drive, printer or other devices as well.

Dick: In other words I can operate or run any application on any other computer while I'm sitting comfortably in my own chair?

Jack: Exactly.

Dick: But what if he doesn't want others to have access to his computer?

Jack: He can easily achieve this through password restrictions.

Dick: Good idea. Thank you for your introduction, Mr. Jack.

Jack: You are welcome.

Related Expressions

1. A network is a collection of computers that are connected together. 网络就是指相互连接起来的一批计算机。

2. You can use the Internet to send E-mail messages or to download files. 你可以利用互联网发送电子邮件，下载文件。

3. Can you tell me your E-mail address? 告诉我你的电子邮箱地址吗？

4. Can you tell me how to search for the songs on the Internet? 你能告诉我怎样在网上搜索歌曲吗？

Dialogue 6   Office Automation System

A: Look, there is a report about office automation system. [1]

B: Office automation system? Oh, it's the first time I have ever heard of it. What is that?

A: It's said that Office automation is the application of computer and co-communications technology to improve the productivity of clerical and managerial office workers. [2]

B: What is the main function of this system?

A: The major functional components of an office automation system include text processing, electronic mail, information storage and retrieval, personal assistance feature, and task management. And these may be implemented on various types of hardware and usually include a video display terminal, a keyboard for input, and a hard-copy output device for letter-quality printing.

B: It sounds very advanced. And who will use it?

A: These were mainly developed to do word processing and record processing, so initially, [3] systems sold by manufacturers were aimed at clerical and secretarial personnel. But more recently, attention has also been focused on systems which directly support the principals-managers and professional workers. Today's organizations have a wide variety of office automation hardware and software components at their disposal.

B: You mean that this system can make the office work easier?

A: Yes, because such systems emphasize the managerial communications function. Electronic mail and filing permit a user to compose and transmit a message on an office automation system. So it makes officer worker's work become easier.

B: Oh, it's very good. I will suggest our manager to use this system.

A: That's a good idea!

Notes

[1] Office automation system：办公自动化系统。

[2] improve the productivity of clerical and managerial office workers：提高办事人员和管理人员的工作效率。

[3] initially： *adv.* 开始，最初。

Dialogue 7   Could You Tell Me What is a Word Processor?

Visitor: Could you tell me what is a word processor?

Secretary: A word processor is like a typewriter with a build-eraser, copier, scissors and glue.

We can edit, replace or move the text after it is typed in.

Visitor: Can you show me how to operate a word processor?

Secretary: Certainly. Turn the computer on, then put the program disk in drive I and the work disk in drive II. Load the program first and name the document you wish to work on.

Visitor: I wonder whether I can manage it.

Secretary: Never mind. There is a help menu in almost every program. If you need help, the menu will tell you what to do next.

Visitor: Oh, does your company use Chinese word processor?

Secretary: Yes, if you use Chinese word processor, you have to be familiar with the coding system for Chinese characters. Here we use "Five-stroke Coding System".

Visitor: I have heard of it.

Secretary: This kind of coding system is very efficient, but it needs some training before one can type Chinese quickly.

Visitor: What other facilities for office automation do you have in your office?

Secretary: Most departments in our company haveduplicators besides computers. Some departments have Fax and telex installations.

Visitor: I see. Thank you very much.

Related Expressions

1. I have to repair some confidential data in my Windows partition. 我要修复存在 Windows 分区上的一些重要数据。

2. We want to have our network system upgraded. 我们想升级网络系统。

3. Is it difficult to maintain them after the upgrade? 升级后的维护是不是很复杂？

4. Format the primary partition C with installer. 用安装程序格式化主分区 C。

5. I advise you to consult your User's Manual in the process because some steps can be very dangerous. 建议您在安装过程中参考用户手册，因为有些步骤可能会很危险。

Dialogue 8 Interview Zhang Xiaoming

Applicant: Good morning. My name's Zhang Xiaoming. I've come here for an interview.

Interviewer: Please take a seat. I'm David Smith from Personnel.

Applicant: Nice to meet you.

Interviewer: Let's begin. What is your major?

Applicant: My major is Electronics & Information Engineering.

Interviewer: Which university are you attending?

Applicant: I am attending Sun Yat-sen University.

Interviewer: How did you get on with your studies?

Applicant: I did well at university.

Interviewer: You speak English fluently. Have you passed CET-4?

Applicant: Yes, I have also got a Cambridge Business English Certificate.

Interviewer: Are you good at computer?

Applicant: Yes, I minored in Computer Application in college.

Interviewer: Thank you forattending this interview. We'll let you know the result as soon as the decision is made.

Applicant: Thank you for your time. I'll be looking forward to your call.

Related Expressions

1. Would you tell me what educational background you have? 请谈谈你的教育背景。
2. How were your scores at college? 你在大学的成绩怎么样?
3. What certificates of technical qualifications have you received? 你有什么技术资格证书?
4. I got the college scholarship in the 2013—2014 academic year. 我获得了2013~2014学年的高校奖学金。
5. My favourate subject was electronic and communication engineering. 我最喜欢的学科是电子与通信工程。
6. I'm proficient in both written and spoken English. 我的英语口语和写作能力都很好。
7. How long is the trial period? 试用期多久?

Dialogue 9 Interview Zhao Zhiming.

Applicant: Good afternoon, sir.

Interviewer: Good afternoon. Take a seat, please.

Applicant: Thank you.

Interviewer: May I have you name, please?

Applicant: My name is Zhao Zhiming.

Interviewer: Tell me about yourself.

Applicant: Sure. I was born in Nanjing, Jiangsu Province and I am 20 years old. I'll graduate from Nanjing College of Information Technology this summer.

Interviewer: What is your major and what are the main courses you've learnt?

Applicant: I'm majoring in Electronic Information Engineering, and I've learnt Fundamentals of Electrical Circuits, Fundamentals of Radio Technology, Analogue Electronic Circuits, Digital Circuits, CAD in Electronics, Measurement and Instrumentation, Network and Communications, OA Equipment, Principle of Color TV etc. And, during the three-year study, I have always belonged to the top in class.

Interviewer: Why are you interested in working for our company?

Applicant: Your company enjoys good reputation in this line. What's more, you offer good opportunities to young people.

Interviewer: Yes, we do. How is your English?

Applicant: I like English and I spend a lot of my spare time studying English. Actually I've passed the Practical English Test for Colleges.

Interviewer: Are you good at computer?

Applicant: Yes, I do well in the related subjects and I've got the certificate of Computer Stage 2.

Interviewer: Thank you forbeing interviewed. We'll let you know the result as soon as the decision is made. Hope to see you again.

Applicant: Thank you. I'll be looking forward to your call.

Related Expressions

1. What're your qualifications? Can you say something about yourself? 您的资历如何? 您能介绍一下自己的情况吗?
2. I've just got my Ph. D. in engineering from Massachusetts Institute of Technology, but I

had worked for two years before I started to work on a doctor's degree. 我刚从麻省理工学院获得工程学博士学位，在攻读博士以前我工作过两年。

3. I've passed the tests for business skills in computer and shorthand. 我通过了计算机和速计的商业技能考试。

4. Are you learning English for any particular reason? 你学英语有没有什么特别的原因？

5. At home in China I work in advertising and many of the words we use are English, so I want to understand English better because that will help my job. 在中国我是做广告的，有很多情况下要用到英语，所以我想提高英语水平，把工作做得更好。

6. Your company enjoys good reputation in this line. 你们公司在这一行业享有盛誉。

Dialogue 10　I'd Like to Book a Booth[1]

J: Hello, Miss Zhao, this is William Johnson.

Z: Hello, Mr. Johnson, nice to hear your voice again. Is there anything I can do for you?

J: Oh, Miss Zhao, after careful consideration, our company has decided to take part in the 5th Guangzhou Hi-Tech Fair.[2]

Z: I'm so happy to hear that. You won't be disappointed by your decision.

J: I'd like to booth. Are there any left?

Z: Yes, but what kind of booth do you prefer?

J: Are there any booths left in Region C?

Z: Sorry, all the booths of Region C have been ordered.

J: Oh, what a pity! Then could you recommend some to me in other regions?

Z: There is one in Region B I'd like to recommend. It's in the middle of the first line facing the gate. It's the best place for a booth.

J: That's great. What is the price for it?

Z: The price of it is 5000 yuan.

J: Fine, I'll take this booth. I'll fax the registration form to you at once.

Z: Thank you!

Notes

[1] booth: n. 售货棚，摊位，小房间

[2] Hi-Tech Fair: 高新技术成果展销会

Dialogue 11　Welcome to Our Booth

Exhibiter: Good morning, Sir. Welcome to our booth.

Visitor: Good morning, I'm interested in your electronic products. Where are your from?

Exhibiter: We are from Jilin, the Northwest of China. This is the first time we are taking part in this Hi-Tech Fair.

Visitor: Oh, what's this? Can you tell me something about it?

Exhibiter: This is one of our new products, and there's nothing similar on the market yet. No one can match us so far as quality is concerned.[1]

Visitor: Is that so? Let me have a close look at it. Do you have any booklets about it?

Exhibiter: Sure, here is a brochure about our company and the products.

Visitor: Could I have your latest catalogues or something that tells me about your company?[2]

Exhibiter: Yes, please leave your business card. I'll fax all the information you need after the exhibition.

Visitor: Fine, here is my card. Please contact me as early as you can. I think we may be able to work together in the future.

Exhibiter: Ok, We would be glad to start business with you. Thank you for coming.

Notes

[1] No one can match us so far as quality is concerned. 就质量而言，没有任何厂家能和我们相比。

[2] Could I have your latest catalogues or something that tells me about your company? 可以给我一些贵公司最近的商品价格目录表或者一些有关说明资料吗？

Dialogue 12　Marketing Planning

Jane: Now we've decided to go global.[1] What form of entry should we use to get into the US market?

Jack: Well, as you saw in my strategy plan, I think that a joint venture would be the best choice. What do you think?

Jane: I think for our type of consumer products a joint venture is not suitable because we don't necessarily need a local partner.

Jack: But their local market knowledge could be invaluable.

Jane: But there are easier ways to gain that local knowledge.

Jack: So what is your suggestion then?

Jane: I think we should acquire the services of a distribution agent as well as a business consultant.[2] These are relatively inexpensive ways to gain local knowledge.

Notes

[1] Now we've decided to go global. 我们已经决定了走全球化路线。

[2] I think we should acquire the services of a distribution agent as well as a business consultant. 我认为我们应该请一位经销代理人和一名业务顾问。

Dialogue 13　Cooperation Intention

A = Sales Representative　　B = Buyer

A: I'm our sales representative, how do you do, what can I do for you?

B: Our company will buy a batch of computers, as the procurement manager secretary,[1] I want to get to know your product.

A: Our company engaged in import and export trade for 5 years, has many professional and qualified partners. Company in good standing, developed many long-term partners, look forward to working with you.[2]

B: I want to know more about your company's products, I hope you can provide me with this. Believe that through the cooperation with your company, we will expand market share in China, China's consumer demand is very strong.

A: I should be very happy to give you any further information you need on it.

Notes

[1] as the procurement manager secretary. 作为采购经理的秘书

167

[2] Company in good standing, developed many long-term partners, look forward to working with you. 公司信誉良好，发展了很多长期合作伙伴，期待与你们的合作。

Dialogue 14　Signing the Contract

Tom: We've brought the draft of our contract. Please have a look.
Mike: How long will the contract last?[1]
Tom: This contract is valid for one year.
Mike: I'm afraid that one year is too short. This contract must be valid for at least three years.
Tom: If everything's going satisfactorily, it could be extended for two years. [2]
Mike: All right. We accept your suggestion.
Tom: Is there any other question?
Mike: No. Nothing more.
Tom: Now, please countersign it.
Mike: Congratulations.

Notes

[1] How long will the contract last? 该合同的有效期是多长？

[2] If everything's going satisfactorily, it could be extended for two years. 如果一切进行得令人满意，可以再延续两年。

Dialogue 15　Telecommunication After-Sale Service

IT Clerk: China Netcom Service. May I help you?
User: Yes. The connection was not good. There was a lot of echo and I couldn't hear well...
IT Clerk: I am sorry, I couldn't hear you. Can you speak a litter louder, please?
User: Something has gone wrong with my...
IT Clerk: China Netcom Service. Can I help you?
User: It's me again. I was disconnected just now when I was dialing to report you the trouble with my fixed phone. So this time I'm using a mobile phone to call you.
IT Clerk: I recognized your trouble just now and I'm sorry for it. Is 475-3233 your fixed phone number?
User: That's right. Something has gone wrong with it.
IT Clerk: Noisy and easily disconnected, isn't it?
User: Yes.
IT Clerk: We'll identify the problem and solve it as soon as possible. How can we contact you?
User: I'm available at 136-7289-9821.
IT Clerk: 136-7289-9821. Is it correct?
User: That's right.
IT Clerk: Goodbye, sir.

Related Expressions

1. I want to use the call display service. 我想办理来电显示业务。
2. You have to pay 6 yuan per month for the service. 此业务每月收6元服务费。
3. 114 directory assistance. May I help you? 114电话导航，能为您服务吗？
4. You can dial the 160 Manual Counter and a technician will answer your question. 您可以拨打160人工服务台，由服务员为你解答疑问。

Appendix B Translation and Keys to the Part of Exercises
(参考译文与部分习题答案)

第1章 电子技术基础

第1单元 电流、电压和电阻

电路的主要作用是沿着特定路径移动或传送电荷，这种电荷的运动形成电流。电流用字母 I 表示，它是电荷随时间的变化率，可表示为 $I = dq/dt$。测量电流的基本单位是安[培] (A)。1A 电流的定义是在 1s 内 1C (6.28×10^{18} 电子) 的电量通过导体的任何一点时的电流。当需要表示比安小的电流量时，可以用毫安 (mA) 和微安 (μA) 作单位。$1mA = 10^{-3} A$, $1μA = 10^{-6} A$。

将单位电荷 (+1C) 从元件的一端移动到另一端所做的功定义为元件两端的电压。电压这个术语 (用字母符号 U 表示) 通常用来表示电位差或电动势，度量电压的单位是伏[特] (V)。使 1A 电流流过电阻为 1Ω 的导体所需的电动势定义为 1V。此外，较小或较大数值的电压可以用毫伏 (mV)、微伏 (μV) 或千伏 (kV) 来表示。

电阻器阻碍电流的流动，比如与发光二极管 (LED) 串联的电阻器就限制了流过 LED 的电流。电阻器的值叫作电阻，用字母符号 R 表示。电阻用欧[姆]来度量，欧[姆]的符号是 Ω。1Ω 的定义是 1V 的电压施加在导体上产生了 1A 的电流时该导体的电阻值的电阻。1Ω 的电阻很小，所以电阻值常常用千欧 (kΩ) 和兆欧 (MΩ) 来表示。

欧姆定律是电子学最基本的定律之一，用来表示导体中电流、电压和电阻之间的关系。该关系用公式表示为"电流 = 电压/电阻"。欧姆定律说明导线两端的电压和流过它的电流的比率等于导线的电阻。常用来描述欧姆定律的数学表达式为"$I = U/R$""$U = IR$"或"$R = U/I$"。

阅读材料 **导体、绝缘体和半导体**

任何允许电子自由流过的物质都叫作导体。一般来说，金属是良导体。良导体的原子结构间存在确定的关系。在良导体中，外层电子还称为原子价电子，可以挣脱自身的轨道进行相对灵活的运动。其原子外环带有 1 个、2 个或 3 个电子，所以大多数金属都是良导体。

阻止电子通过的物质叫作绝缘体。绝缘体外层几乎没有易于移动的电子。没有理想的绝缘体：首先因为杂质 (不同材料) 不可能完全去除；其次很小的热量即可使原子价电子挣脱原子的束缚成为自由电子。

通常绝缘体的原子结构很稳定，其中最典型的是外层有 4 个电子的结构。这种结构中没有易于移动的电子。好的绝缘体，比如碳的化合物和钻石，它们有类似的原子结构。

半导体是一系列导电性能较差的物质，它们既不属于导体也不属于绝缘体。一般来说，半导体与绝缘体的主要区别在于它们的外层电子比绝缘体更易于从轨道上自行分离出来。典型的半导体材料是锗和硅。

将杂质掺入纯净的半导体中就形成了半导体材料。半导体材料中要么会产生多余的自由电子，要么轨道中有电子空缺。产生多余的电子时，称之为 N 型半导体材料；出现轨道上的电子空缺时，称之为 P 型半导体材料。N 型和 P 型半导体都是由加工过的材料制成的。在半导体中掺入杂质的过程叫作掺杂。

Keys to Exercises of the Text

Ⅱ. 1. T 2. F 3. F 4. F 5. T 6. T

Ⅲ. 1. C 2. C 3. C 4. A 5. B 6. B 7. A 8. A

Ⅳ.

1. The term voltage (represented by the letter symbol V) is commonly used to indicate both a potential difference and an electromotive force.

2. The magnitude/ value/ amount of resistor is called resistance.

3. To move or transfer charges along specified paths constitutes/ makes up an electric current.

4. "Ohm's Law" is one of the fundamental laws of electronics.

Keys to Exercises of the Reading

Ⅱ. 1. 许多原子价电子　　　　　2. 可挣脱轨道进行相对灵活的运动
　 3. 外层有4个电子的结构　　4. 既不属于导体也不属于绝缘体
　 5. 或者会产生多余的自由电子，或者轨道中有电子空缺

第2单元　电子元器件

这些是绝大多数电子元器件的表示符号。能够识别更多的普通元器件并掌握它们的实际用途是很重要的。下面画出了一些电子元件，值得注意的是，一种类型的元器件往往可以用几种不同的符号表示。

1. 电阻器

电阻器阻碍电流的流动。比如，一个电阻器可以与一个发光二极管 LED 串联来限制流过 LED 的电流。图2-1 表示电阻器实物图及其在电路图中的符号。电阻器可以连接在任一回路中。它们不会因焊接产生的高温而受到损坏。

2. 电容器

电容器存储电荷。由于电容器使交流信号容易通过，但阻止直流信号，因此它们经常应用于滤波电路中。图2-2 为电容器实物图及其在电路图中的符号。

3. 电感器

电感器为无源电子元件。它以磁场的形式存储能量。电感器就是一个将导线缠绕了若干匝数的线圈，通常会把它缠绕在像铁心这样的磁性材料上。图2-3 表示电感器实物图及其电路符号。

4. 二极管

二极管允许电流只从一个方向流过。电路符号的箭头表示电流能够流过的方向。二极管是真空管的电子版。实际上，早期的二极管就称为真空管。图2-4 是二极管实物图及其电路符号。

5. 晶体管

标准晶体管分为两类，NPN 型和 PNP 型。它们在电路图中的符号不同。字母（N 和 P）表示制造晶体管的半导体材料不同。图2-5 是晶体管实物图及其电路符号。

6. 集成电路（芯片）

集成电路通常称为 IC 或芯片。它们将复杂电路固化在小型半导体（硅）芯片上。该芯片被封装在一个塑料固定物上，引脚间隔距离为 0.1in（2.54mm），这样的栅格将适合带形板和面包板的孔距。在封装里面用很纤细的导线连接芯片引脚。图2-6 是集成电路实物图。

7. 发光二极管（LED）

电流通过 LED 时，会使 LED 发光。

LEDs 必须连接正确的回路，电路图中可用"a"或者"+"表示阳极，用"k"或者"-"表示阴极。阴极是短的引脚，并且在 LEDs 圆形体内可能是细小扁平的那端。图 2-7 表示 LED 实物图及其电路符号。

8. 其他电子元件

图 2-8 表示其他电子元件的实物图和电路符号。

可变电阻器、蜂鸣器、扬声器、熔丝、灯泡/灯丝、电动机、螺线管、开关。

Keys to Exercises of the Text

Ⅰ. 1. T 2. F 3. T 4. T 5. T 6. T 7. T

Ⅱ. 1. d 2. e 3. a 4. c 5. b

Ⅲ. 1. Resistor 2. filter 3. version 4. plastic 5. only one

Ⅳ.

1. It is essential/ necessary that you can recognize the more electronic components and grasp what they actually do.

2. Resistors are not damaged by heat when soldering.

3. Metals do not melt until heated to a definite temperature.

4. The field of electronics includes the electron tube, transistor, integrated circuit and so on.

第 3 单元　晶体管及其基本电路

晶体管是现代电子学中最重要的器件。它不仅可以用作分立元件，而且在集成电路芯片中可包含成千上万的晶体管。

晶体管是三端器件，可用作放大器和开关，晶体管有以下两种基本类型。

1）结型晶体管（通常称为晶体管）：其工作依赖于两种载流子，即多数载流子和少数载流子的流动，并有两个 PN 结。

2）场效应晶体管（简称为场效应管）：其电流仅由多数载流子提供（可以是电子，也可以是空穴），且只有一个 PN 结。

晶体管由三层半导体材料构成：一种类型的薄层在中间，两边分别是另一种类型的材料。有两种安排：N 型在中间 P 型在两边（PNP）；P 型在中间 N 型在两边（NPN）。中间称为基极，两边分别为发射极和集电极（如图 3-1 所示）。

晶体管是调节流过它的电流的电子器件，电流从电源流进发射极，穿过很薄的基区，再从集电极流出，电流始终朝着这个方向流动。改变基极电流，这个电流的大小也会随着改变，基极电流只需很小的变化，就会引起集电极电流很大的变化。正是这种能力才使晶体管起到放大作用。

晶体管在电路中有三种基本连接方式：共基极、共发射极和共集电极（如图 3-2 所示）。在共基极连接中，信号从射-基回路输入，从集-基回路输出。因为射-基回路的输入阻抗较低，约为 0.5~50Ω，而集-基回路的输出阻抗较高，约为 1000Ω~1MΩ，因此，采用这种结构的电路，其电压或功率增益可达 1500 倍。

在共发射极连接中，信号从基-射回路输入，从集-射回路输出。与共基极连接相比，这种结构有中等的输入和输出阻抗，输入电阻约为 20~5000Ω，输出电阻约为 50~50000Ω。因为既有电压增益又有电流增益，这种结构的功率增益可以达到大约 10000 倍（约 40dB）。

第三种连接方式是共集电极连接。在这种结构中，信号从基-集回路输入，从射-集回路输出。因为晶体管的输入阻抗高而输出阻抗低,所以电路的电压增益小于1,而功率增益比共发射极或共基极连接的要低。

Keys to Exercises of the Text

Ⅱ. 1．B 2．C 3．B 4．C 5．A 6．C

Ⅲ

1. Transistors are the most important device in electronics.

2. There are two basic types: 1) the junction transistor (usually called the transistor); 2) the field effect transistor (called the FET).

3. A transistor is an electronically device that regulates the current flowing through it.

4. Because the input impedance of the transistor is high and the output impedance low in this connection, the voltage gain is less than one and the power gain is usually lower than that obtained in a common-base or a common-emitter circuit.

阅读材料　　　　　　　　　　　　**尼古拉·特斯拉**

尼古拉·特斯拉于 1856 年出生于克罗地亚的斯米良利卡。他是一个西伯利亚东正教神父的儿子。特斯拉在奥地利工业学校学习工程学。他在布达佩斯做电气工程师,后来在 1884 年移民到美国后在爱迪生机器公司工作。他与 1943 年 1 月 7 日在纽约去世。

在他的一生中,特斯拉发明了荧光灯、特斯拉感应电机、特斯拉线圈、交流电供电系统,这个系统包括一个电机和变压器和三相电。

在最高法院于 1943 年推翻了古格列尔莫·马可尼的专利而判特斯拉早期专利有效后,特斯拉现在也被公认发明了现代无线电;当一个叫奥的斯·庞德的工程师谈论马克尼的无线电系统时对特斯拉这样说"好像马可尼在您的基础上起跳",特斯拉答道,"马克尼是个不错的伙计。让他接着干吧。他使用了我的 17 件专利。"

特斯拉在 1891 年发明的特斯拉线圈,现在还使用在电视机和其他电子产品上。

尼古拉·特斯拉——神秘的发明

在发明了生产交流电的方法并申请专利 10 年后,尼古拉·特斯拉声称发明了一种不需要任何燃料的发电机。这个发明已经在公众面前丢失了。特斯拉声称他的这个发明可以利用宇宙射线来驱动动力装置。

特斯拉被授予了超过一百件专利,还有很多数不清的未申请专利的发明。

尼古拉·特斯拉和乔治·威斯汀豪斯

1885 年,威斯汀豪斯电气公司的老板乔治·威斯汀豪斯购买了特斯拉的发电机、变压器和电动机的专利权。威斯汀豪斯使用了特斯拉交流电系统来给 1893 年在芝加哥召开的哥伦布纪念博览会提供照明。

尼古拉·特斯拉和汤姆逊·爱迪生

尼古拉·特斯拉是汤姆逊·爱迪生在 19 世纪末的竞争对手。实际上他在 19 世纪 90 年代比汤姆逊·爱迪生更出名。他的多相电发明为他在世界上赢得了很高的知名度和大量财富。在他的人生顶峰时期他是很多成功人士的密友,这些人包括诗人、科学家、工业家和银行家。然而特斯拉去世时贫困潦倒,丢失了财富和在科学界的声誉。在他从声名远扬到变得默默无闻的下降过程中,特斯拉留下了名副其实的这些发明遗产给后人,还有现在人们仍为之着迷的一些预言。

Keys to Exercises of the Reading

Ⅰ. 1．T 2．F 3．T 4．T

II.

1. Tesla claimed that he worked from 3 a. m. to 11 p. m., no Sundays or holidays excepted.

2. After 1890 Tesla experimented to transmit power generated with his Tesla coil by inductive and capacitive coupling with high AC voltages.

3. Trying to come up with a better way to generate alternating current Tesla developed asteam powered reciprocating electricity generator which he patented in 1893 and introduced at the Worlds Colombian Souvenir Exposition that year.

第 4 单元　集成电路

集成电路是几种相互连接的电路元器件，如晶体管、二极管、电容器和电阻器等的组合。它是用半导体材料制成的小型电子器件。第一块集成电路是在 20 世纪 50 年代由德州仪器的 Jack、Kilby 和 Fairchild 半导体公司的 Robert Noyce 合作开发的。

电学上把组成集成电路的彼此相连接的元器件称为集成元器件。如果集成电路只包含一种类型的元器件，便称为元器件组件。

在各种设备包括微处理器、音频和视频设备以及汽车中都要用到集成电路（见图4-1）。集成电路通常根据其包含的晶体管和其他电路元件的数量来归类。

· SSI（小规模集成电路）：每个芯片中有 100 个以下的电子元器件。

· MSI（中规模集成电路）：每个芯片中有 100~3000 个的电子元器件。

· LSI（大规模集成电路）：每个芯片中有 3000~100000 个的电子元器件。

· VLSI（超大规模集成电路）：每个芯片中有 100000~1000000 个的电子元器件。

· ULSI（甚大规模集成电路）：每个芯片中有 100 万个以上的电子元器件。

随着在一个芯片上集成大量晶体管的能力（即集成电路的集成量）的提高，对专用集成电路的需求已更加普遍。至少对大批量的应用来说更需要专用的集成电路。硅芯片技术的发展使得集成电路设计者可以在一个芯片上集成几百万个以上的晶体管，甚至现在可以在单一芯片上集成一个中等复杂的系统。

集成电路的发明是电子工业的一次重要革命。凭借这一技术，完全可以实现尺寸缩小和重量减轻，更重要的是能够实现高可靠性、良好的工作性能、低成本和低功耗。集成电路已广泛应用在电子工业中。

Keys to Exercises of the Text

II. 1. T　2. T　3. F　4. T　5. F　6. F

III.

1. An integrated circuit (IC) is a combination of a few interconnected circuit elements on a piece of semiconductor chip.

2. The electrically interconnected components that make up an IC are called integrated elements.

3. ICs are widely used in the electronic industry.

4. Integrated circuits are often classified by the number of transistors and other electronic components

IV.

集成电路看起来只不过是一个微小的金属片，或许一边只有 1/2 厘米长，比一张纸厚不了多少。芯片这么小，如果它掉到地板上，就很容易跟灰尘一起扫走。尽管它很小，芯片的生产过程却应用了当代最先进电子技术。到今天的发展水平，芯片可能包含一万多甚至数百万独立的电子元器件（这些电子元器件有许多不同的功能，诸如二极管，晶体管，电容器和电阻器等）。

第5单元　电子仪器

1. 万用表

万用表是一种通用仪表，可用来测量直流和交流电压、电流、电阻，有的还能测量分贝（放大倍数的情况）。有两种万用表：一种是用指针在标准刻度上的移动来指示测量值的模拟万用表（见图5-1a），另一种是用电子数字显示器显示测量值的数字万用表（见图5-1b）。这两种万用表都有一个正极（+）插孔和一个公共端（-）插孔用来插入测试笔，一个功能选择开关用来选择测量对象：直流电压、交流电压、直流电流、交流电流或电阻，一个范围选择开关用来选择范围以测出精确读数。万用表还可能有其他插孔用来测量高电压（1~5kV）和大电流（高达10A）。对一些特殊的万用表而言，还会有一些其他功能上的变化。

除了功能选择开关和范围选择开关（有时它们合并为一个开关），模拟万用表可能还有一个极性开关，可以很方便地交换测试笔的极性。指针常常有一个旋钮来机械调整。另外在测量电阻时有一个零点调节控制钮用来对万用表内部的电池电压不足做补偿调节（即保证电阻为0时指针指向零值）。模拟万用表可以测量正电压和负电压，只需要简单地对调一下两个测试笔或拨一下极性开关。数字万用表通常会自动在显示器上指示出极性。

为了确保读数正确，万用表必须与电路正确连接。一个电压表（万用表测量电压时）应与被测电路或元器件并联。当测量电流时，电路必须断开，插入万用表表笔，使万用表与被测电路或元器件相串联。当测量电路中局部电路（或元器件）的（等效）电阻时，必须关掉电路中的电源，使万用表与该局部电路（或元器件）并联。

2. 示波器

示波器（见图5-2）是一个图像显示设备，它显示一个电子信号的图像。当信号输入到示波器中时，就产生一个电子束，该电子束被聚焦、加速并适当偏离，在阴极射线管的显示屏上显示电压的波形。

示波器通常显示电压信号如何随时间变化：其纵轴 Y 表示电压，横轴 X 表示时间。示波器屏幕上电压波形的幅度可以通过数出电压波峰与波谷之间的纵向距离来确定（见图5-3）。将这个值乘上 V/cm 控制钮的设定值就得到电压的幅值。比如说，假若电压的峰-峰值幅度为5cm，控制钮设在1V/cm处，那么，峰-峰值电压为5V。

用示波器的水平标尺可以测量时间值。时间测量包括测量信号的周期、脉冲宽度和频率。频率是周期的倒数，所以一旦知道了周期，频率就是用1除以周期即可得到。

一个波形的频率可以通过在水平方向数出该波形一个周期的长度来确定，将这个值乘上 t/cm 控制钮的设定值就得到它的周期。例如，假定一个波形长4cm，控制钮设在1ms/cm，那么周期是4ms，则频率可以用下面的公式算出：

$$f = \frac{1}{T} = \frac{1}{4\text{ms}} = 250\text{Hz}$$

假定控制钮设在100ms/cm，则周期是400ms，频率为2.5kHz。

双踪示波器具有同时显示输入信号和输出信号的优点，可以显示输出信号是否有失真，并表明输入/输出信号的相位关系，即将两路信号的波形重叠在一起可以较好地显示出输入与输出信号相位的漂移。

Keys to Exercises of the Text

II.

1. 通用仪表　　　　　　　2. 交换测试笔

3. 机械调节　　　　　4. 电压振幅
5. 双踪示波器　　　　6. 信号发生器
7. 模拟万用表　　　　8. 相位关系
9. 显示电压波形　　　10. 正电压

Ⅲ

1. The analog multimeter may have a polarity switch to facilitate reversing the test leads.

2. There are some variations to the functions used for specific meters.

3. The oscilloscope displays/ draws a graph of an electrical signal.

4. A dual-trace oscilloscope is advantageous to show the input signal and output signal of one circuit in the same time.

5. The two traces may be placed over each other (superimposed) toindicate better the phase shift between two signals.

阅读材料　　　　　　　如何使用测试仪表？

该仪表是设计用来测量直流电压、交流电压、电阻、导电性能和测试二极管的。该仪表具有 $3\left(\frac{1}{2}\right)$ 位数字的液晶显示，因此它也被称作数字多用表。

该仪表的使用方法示例如下：

1. 直流电压的测量（见图5-4）

1）功能开关置于 DCV。

2）测试表笔接在被测电路上。

3）读显示值。

注：

1）表笔极性反接时，会出现"－"（负号）。

2）在测量包含有尖锋脉冲的电压时（如电视中的水平输出信号）使用正接的测试表笔。

2. 交流电压的测量（见图5-5）

1）功能开关置于 ACV。

2）测试表笔接在被测电路上。

3）读显示值。

注：不用考虑测试表笔的极性。

3. 电阻测量（见图5-6）

1）功能开关置于 Ω 档。

2）测试表笔接在被测电路上。

3）读显示值。

注：接表笔之前要确认已断开被测电路的电源。

4. 导电性能测试（见图5-7）

1）功能开关置于符号为 " ·)) " 上。

2）表笔接在被测试电路上。

3）有蜂鸣音且显示标识 " ·)) " 时表示导通良好。

5. 二极管测试（见图5-8）

1）功能开关置于符号为 " ·)) " 上。

2) 对正接的二极管, 当黑表笔接其阴极, 红表笔接阳极时, 显示为正向电阻值; 当测试表笔接反时, 显示数字为 "1."。

3) 当测试表笔开路时, 显示读数为 "1."。

Keys to Exercises of the Reading

Ⅰ.

1. This instrument is designed to use for measuring voltages, measuring resistance, conductivity test and diode test.

2. The following is illustrating its operation.

3. Minus sign " – " is displayed when the polarity of the test leads is reversed.

4. Connect the test leads to the circuit to be measured.

第 6 单元　便携式媒体播放器

便携式媒体播放器 (PMP) 是一个能够存储和播放数字媒体如音频、图像、视频、文件等的消费性电子装置。数据通常存储在硬盘驱动器、微驱动器或闪存中 (如图 6-1 所示)。

数字音频播放器通常按照存储介质来分类, 可分为:

- 基于 Flash 的播放器: 数字音频文件储存在内部闪存中的非机械固态设备。因为它们是静态存储装置, 并没有移动的部件, 因而需要的电量更少, 在播放过程中很少会出现跳读现象, 而且与硬盘存储播放器相比, 更能经受住诸如掉落或碎裂等此类情况所造成的损害。

- 基于硬盘的播放器或数字点唱机: 这种播放器从硬盘驱动器 (HDD) 读取数字音频文件。在 2010 年, 它就有高达 500GB 的容量。按照典型的编码率, 意味着上万首歌曲可以存储在一个播放器里。其缺点是硬盘驱动器消耗更多的电力, 播放器较大和较重, 与固态存储相比更容易损坏。

- MP3 CD/DVD 播放机: 便携式 CD 播放机可以解码和播放存储在光盘上的 MP3 音频文件。这样的播放器通常比硬盘或闪存播放器便宜得多。它们使用的空白的 CD-R 非常便宜, 一般每张光盘花费不到 0.15 美元。这些器件具有能够播放标准的 "红皮书" CD-DA 音频 CD 的功能。它的缺点是, 因为这些设备的旋转磁盘速度很低, 如果它们在播放过程中遇到不规则的加速度 (震动), 就更容易出现跳读或误读文件的情况。该播放器比硬盘型播放器坚固, 一般不容易因摔到地上而出现永久性损坏。由于 CD 通常只能容纳大约 700MB 的数据, 一个大型文库往往需要多个磁盘来容纳。然而, 一些比较高端的播放机也能读取和播放那些存储在容量更大的 DVD 中的文件; 有的播放机还能播放和显示视频内容如电影。我们还需了解的是, 这些设备具有较大的宽度, 因为它们要容纳整个 CD 光盘。

- 网络音频播放器: 播放器通过联网 (WiFi) 接收和播放音频。这种类型的播放器一般没有自己的存储器, 必须依靠一台服务器, 通常为一台连接在同网络上的个人计算机, 来提供音频文件进行播放。

- USB 主机/存储卡音频播放器: 播放器依靠 USB 闪存驱动器或其他存储卡读取数据。

PMP 能够播放数字音频、图像和视频。通常使用彩色液晶显示屏 (LCD) 或者有机发光二极管 (OLED) 屏幕进行显示。不同类型的播放器包括的功能有: 录制视频, 通常借助可选附件或电缆; 录制音频, 借助一个内置麦克风或线路输出电缆或 FM 调谐器。有些播放器具有存储卡读取器, 鼓励用户将外部存储器或传输介质装在播放器上。

Keys to Exercises of the Text

Ⅱ．display　record　microphone　readers

Ⅲ．1．B　2．E　3．D　4．C　5．A

Ⅳ．

1．自从我们奉行改革开放政策以来，我们的工业飞速发展。

2．数字音频播放器通常按照存储介质来分类。

3．PMP 能够播放数字音频、图像和视频。

4．你可以对找到的内容进行分类并加标签、设为私有、分享给朋友或设为公开。

第 2 章　通　信　技　术

第 7 单元　光 纤 通 信

广义地说，把信息从一个地方传递到另一个地方就称为通信。当信息跨越一段距离被传送时，就需要一个通信系统。在通信系统中，信息传送时常是通过把信息叠加在电磁波上或对电磁波进行调制来实现的，这样的电磁波起着传送信号的作用。经过调制的载波随后被传送到要到达的目的地，在那里被接收并且通过解调还原成原始信息。在运用电磁载波的领域，虽然用无线电波、微波以及毫米波的频率为载波频率的高新技术已很成熟，但仍可选择光波频率段的电磁波作为载波来实现通信。

典型的光纤通信系统如图 7-1 所示。信号源提供电信号给发射机，发射机组成一个电路平台来驱动光源以完成对光载频率的调制。光源是发光二极管或是半导体激光管，它完成光电转换。传输媒介由光纤组成。接收机包括一个含光检测器的电路驱动平台，用以完成对已调光载波的解调。光检测器如光敏二极管、发光晶体管以及光敏电阻，用于检测光信号，进行光电转换。因此，在光系统链路的两端都需要有电接口，并且在现阶段，信号处理通常是通过电路实现的。

模拟或数字的信号均可用来调制光载波。模拟调制是指从光源处发射的连续光强度的变化，而数字调制则不然，它获得光强度离散的变化（如开关脉冲）。虽然实施起来更简单，但模拟调制在光调制系统中效率较低，与数字调制相比，需要高得多的信噪比。同时，尤其是在高频调制中，半导体光源不总能提供模拟调制所必需的线性。基于上述原因，与数字光系统相比，模拟通信链路通常被限制使用在较短的通信距离和较窄的带宽上。

为便于光传输，信源的数字信号首先要被适当地编码。激光器的驱动电路通过这些已编码的数字信号来直接调制激光器的发光强度。然后数字光信号被注入光纤。在接收端，信号通过雪崩光敏二极管检波器（APD）后进入前置放大器和均衡器或滤波器，用来提供增益，对信号进行线性处理和减少噪声带宽。最后，解码得到原始信号。

Keys to Exercises of the Text

Ⅱ．1．achieved　2．detection, conversion　3．digital optical signal　4．decoded
5．signal processing

Ⅲ．1．A　2．A　3．A　4．C　5．A　6．A

Ⅳ．

1．Communication may be broadly defined as the transfer of information from one point to another.

2．When the information is to be conveyed over any distance a communication system is usually required.

3．Within a communication system the information transfer is frequently achieved by superimposing or modulating the information on to an electromagnetic wave which acts as a carrier for the information signal.

4. Finally, the signal obtained is decoded to give the original digital information.

阅读材料　　　　　　　　　　　　**无线电通信技术**

　　无线电术语描述的是无线电通信，即用电磁波（而不是其他种类的电线）在局部或全部通信线路上传递信号。19 世纪初，第一批无线电发射器发射了无线电报。之后，由于调制后的无线电可以传递声音和音乐，这种介质就称作了"无线电接收装置，或无线电广播设备"。随着电视、传真、数据通信陆续出现，以及更大面积光谱领域的有效使用，无线电术语再次复兴。

　　无线电的分类如下：
- 固定无线电：指在家庭和办公室操作的无线电装置或系统，尤其指通过专用调制解调器与互联网连接的设备。
- 移动无线电：指移动交通工具（车辆、飞机、船舶）上使用的无线电装置或系统，如汽车上的无线电话和个人通信服务程序（PCS）。
- 手提式无线电：指在办公室、家庭或交通工具之外电池驱动的独立操作无线电装置或系统，如手机和个人通信服务设备。
- 红外线无线电：指用红外线辐射传递数据的装置，用于某一范围内的通信系统和控制系统。

目前常用的无线电设备有：
- 移动电话和传呼机：把便携式和移动式应用程序联为一体，供个人或商务使用。
- 全球定位系统：使车辆驾驶员、船舶上的船长、飞机上的飞行员能够给地球的任何地方定位。
- 无线计算机外围设备：普通的例子有不带电线的鼠标器。键盘和打印机也可以通过无线电与计算机相连。
- 无线电话装置：与移动电话不同，这些装置受一定范围的限制。
- 家庭娱乐电器遥控装置：最常见的例子有录像机遥控器和电视频道遥控器；有些高保真立体声音响系统和调频广播接收器也使用这种技术。
- 车库门遥控开启装置：这种最老式的无线电装置，是消费者的常用装置之一，该装置用无线电频率工作。
- 双向无线电通信装置：包括"业余爱好者和市民无线电服务"设施，以及商业、海军和军用通信系统。
- 婴儿监护器：这些装置是简化了的无线电发送或接收器，在一定范围内使用。
- 卫星电视：使任何地方的电视观众都可以从数百个频道中选择电视节目。
- 无线电局域网：给计算机的商业用户提供了灵活性和可靠性。

无线电技术正在迅速发展，在全世界人们的生活中发挥越来越多的作用。不仅如此，有相当多的人正在直接或间接地依赖着这项技术。无线电通信和控制方面出现了一些更专用和奇妙的用途：
- 全球数字移动电话系统（GSM）：在欧洲和世界其他地方使用的数字移动电话系统，实际上已经成了欧洲无线电电话技术标准。
- 通用包无线电服务程序（GPRS）：一种信息包无线电通信服务，将移动电话和计算机用户与互联网相连。

- 增强型数据全球数字移动电话系统环境（EDGE）：一个更快版本的全球数字移动电话无线电系统。
- 通用移动电信系统（UMTS）：一种宽带信息包系统，该系统给世界任何地方的手提计算机和手机提供一套兼容性服务。

无线电应用协议信号（WAP）：一组使无线电设备标准化的通信协议信号，使手机、无线电收发报机等可以访问互联网。i-型智能电话：日本推出的世界第一个智能电话，该电话装置上可显示颜色和图像，可浏览环球网。

Keys to Exercises of the Reading

Ⅱ.
1. 无线局域网　　　　　　2. 无线电话装置
3. 全球移动通信系统　　　　4. 无线应用协议信号

第8单元　卫星通信

20世纪80年代后期，卫星通信已经成为日常生活的一部分，打国际电话就像给住在同一街区的朋友打本地电话一样简单。同样，我们获悉国际大事，像英国的竞选、法国的网球比赛就像得知本地政治、体育新闻一样平常。这样，电视新闻节目每天晚上都将全世界的声音和画面带进我们的家中。

这种进行全球性信息交流的能力，无论是电话还是新闻转播都是通过一个强有力的通信工具——卫星，才得以实现的。对于我们中那些并非生长在太空时代的人们来讲，卫星通信是人们长期以来一种梦想的顶点，这个梦想可以一直追溯到卫星这个词只是几个天才头脑中灵感的想象那个年代。这些先驱者中就包括亚瑟·克拉克，他在1945年就产生了全球卫星系统的想法，这个想法随后就发展成为一个遍布全球的复杂卫星网络。

亚瑟·克拉克设想的卫星通信系统将导致开发一个相当复杂和麻烦的基于地面的空间站的网络。幸运的是，这个问题随着1963年和1964年同步卫星的发射成功而消失。现在的卫星不是高速地围绕地球运行，而是静止或固定在空中，目前大部分通信卫星都被定位在相对地球静止的轨道或称之为"槽"的位置上。

简单地说，一颗处于与地球旋转同步轨道上的卫星对于地球上部分地区而言似乎是静止不动的，在赤道上空35881km的高度上，卫星与地球以同样的角速度运行，即它的运转与地球的自转是同步的。尽管卫星是以很高的速度运行，对于一个地球上的观察者来讲，它总是停留在天空中的同一个位置上。

位于一个对地面相对静止的轨道上的卫星的主要作用在于它可以一天24h与它覆盖的地面站通信。这个轨道的位置同时使建立卫星与地面站之间的通信链路更加简单化。当地面站的天线处在适当的位置时，在一段很长时间内天线的位置只需作微小的调整。只有当地面站要与另外一个卫星建立联系时，天线的位置才需作显著的调整。在这以前，地面站的天线必须物理跟踪天空中运动的卫星。

基于这些原理，围绕地球等距离位置上放置三颗卫星就可以建立一个全球的通信系统，以使地球上的每一个点都能与卫星相连（如图8-1所示）。这个概念还是以亚瑟·克拉克的最早有关全球通信网络蓝图作基础的。

Keys to Exercises of the Text

Ⅱ. 1. D　2. C　3. B　4. A
Ⅲ. communicate with　　station　　antenna　　track
Ⅳ.

1. 这个想法随后发展成为一个遍布全球的复杂卫星网络。
2. 幸运的是，学习和学会更好的学习，将永远是你个人和职业生活中最宝贵的技能之一。
3. 这是一个没有其他途径可以达到的精彩高潮。
4. 要确定项目的范围和所需的工作，应该在做出任何建议前确定组织的准备情况。

第 9 单元　全球定位系统（GPS）

GPS 指的是全球卫星定位系统。GPS 使用卫星技术，使地面终端能够确定它在地球上的经度和纬度的位置。

接收器同时测量来自三个或三个以上卫星的信号，并通过这些信号的传播时间数据来确定自己的位置（见图 9-1）。

GPS 运行时使用三边测量法。三边测量的过程是这样的：在未知点两侧设定两个或两个以上的已知点，与未知点构成虚拟三角形，通过测量未知点到已知点的边长来确定未知点的位置。

在 GPS 中，两个已知点位置数据由两个全球定位系统卫星提供。这些卫星不断传送一个识别信号。GPS 接收机通过测量每一个信号在 GPS 卫星和 GPS 接收机之间的传播时间来测量它们之间的距离。

全球卫星定位系统分为三个部分：
- 空间部分
- 控制部分
- 用户部分

GPS 系统使用 21 颗运行卫星来运作，轨道中还有另外 3 颗后备卫星。GPS 使用的 NAVSTAR 卫星由罗克韦尔国际公司制造。每一个 NAVSTAR 卫星约 5m 宽（太阳能发电板展开时的宽度），重量大约 900kg。

GPS 卫星的轨道离地球的海拔高度约为 20200km。每个 GPS 卫星绕地球轨道的运转周期为 11h58min，这意味着每一个 GPS 卫星每天绕地球两周。

这 24 颗卫星运行在 6 个轨道面或路径上，这意味着每个轨道面有 4 个 GPS 卫星在运行。6 个轨道面的每两个之间呈 60°隔开。所有这些轨道面都与地球赤道呈 55°角倾斜。

为了 GPS 能够进行跟踪工作，对全球定位系统的访问是必需的，一个全球定位系统接收机也是必不可少的（见图 9-2）。全球定位系统接收机能收到上方绕轨道运行的 GPS 卫星传送的信号。一旦这些卫星传送的信号被全球定位系统接收机收到，位置及其速度和方向等信息就能被计算出来。

接收机含有一个数学模型来计算这些信息的影响，卫星也传播一些相关的信息以帮助接收机正确地估算正确的传播速度。一些延迟源，比如电离层，基于无线电波传播频率的不同，会对不同频率的无线电波的速度有相应的影响，双频率接收机能测量它们对信号的影响。

为了测量卫星和接收机之间的时间延迟，卫星重复发送一个 1023 字节长的伪随机序列；接收机收到后建立一个完全相同的序列并移动它直到两序列匹配。

不同卫星使用不同的序列，这样，当不同的卫星发送相同频率的信号时，接收机能够区别信号来自哪个卫星。这是码分多址技术的一种应用（CDMA）。

有两个频率现在被 GPS 使用：1575.42 MHz（称为 L1 频率）和 1227.60 MHz（L2 频率），L1 信号携带一个公开使用的粗码（C/A）和一个加密码 P（Y）。L2 信号通常只携带

P（Y）码。

Keys to Exercises of the Text

Ⅱ.

1．b 2．e 3．d 4．c 5．f 6．g 7．h 8．a

Ⅲ．

based；navigation；drove；while；justification；deployment；why；enable

Ⅳ．

1．全球定位系统是一个全球导航卫星系统，它在全天候情况下提供地表或近地地点的接收器位置和时间信息，这些地点应该无阻碍地被四个或更多的卫星看到。

2．为了克服以往导航系统的局限，GPS系统于1973年在美国被开发。

3．俄罗斯的全球导航卫星系统与GPS系统同时开发，但直到20世纪中期的时候才解决了覆盖率小的问题。

4．GPS系统的设计部分是基于20世纪40年代早期英国皇家海军开发并用于"二战"的路基雷达导航系统。

阅读材料　　　　　　　　　　视　频　会　议

视频会议是指在两个或两个以上在不同地点的人通过用计算机网络传递声音和视频数据在一起开会。每个参会者都有与计算机相连接的摄像机、话筒和扬声器。当两个参会者对话时，他们的声音通过网络送给对方，摄像头前出现的画面显示在对方的计算机监视器的窗口中。

为使视频会议能够进行，会议参与者必须使用同样的客户软件或相互兼容的软件。许多免费和共享的视频会议工具软件可从网上在线下载。许多网络摄像机也捆挷视频软件。许多较新款的视频会议（软件）包还能与即时通信客户端软件集成，用于多点开会和合作。

最近几年，视频会议已成为教室中远距离通信的一个常用形式，用来提供远程学习，嘉宾演讲和多校合作项目，是一种很有效的节约成本的方式。许多人认为视频会议提供了标准即时通信和电子邮件所不能做到的可视连接和交互作用功能。

第10单元　4G网络技术

4G是目前正在开发的具有高速移动宽带能力的第四代无线通信技术。它的特点是高速数据传输和更好的声音质量。虽然ITU（国际电信联盟）尚未给出确切的定义，行业已经确认了以下的4G技术：

- WiMAX（全球微波互联接入）
- 3GPP LTE（第三代合作伙伴长期演进项目）
- UMB（超移动宽带）
- Flash-OFDM（快速低延迟访问的无缝切换正交频分复用）

正在开发的4G技术可以满足QoS（服务质量技术）和网络流量优先化的速率要求，从而保证良好的服务质量。这些机制对于使用大带宽的应用是必需的（如图10-1所示）。例如：无线宽带上网、MMS（多媒体信息服务）、视频聊天、移动电视、高清晰度电视、DVB（数字视频广播）、实时音频、高速数据传输。

ITU为WIMAX和LTE设定的目标是：当用户相对于基站高速移动时，数据传输速率达到100Mbit/s；位置固定时，速率达到1Gbit/s。

大量的4G兼容设备随着行业的发展出现了。它不仅局限于4G手机或笔记本式计算机，在摄录像机、游戏装置、自动售货机和电冰箱等设备上也有应用。

现在的趋势是给每一个提供和纳入嵌入式 4G 模块的便携式装置提供无线互联网接入。4G 技术不仅可以提供宽带互联网连接，而且具有更高级别的安全性，有利于像自动售货机和计费装置等包含金融交易的设备的安全运行。

在所提到的 4G 技术中可以观察到以下几个主要特征：

- 多输入多输出：通过包括多天线和多用户多入多出的空间处理手段，达到超高的频谱效率。
- 频域均衡，如在下行链路中的多载波调制（正交频分复用）或单载波频。

上行链路中的频域均衡：利用频率选择性信道特性，而且没有复杂的均衡。

- Turbo 原理的纠错码：尽量减少在接收端信噪比的要求。
- 基于无线信道的调度：采用时变信道。
- 链路适配：自适应调制和纠错码。
- 利用移动 IP 的移动性。
- 基于 IP 架构的家庭基站（连接到固定网络宽带基础设施的家庭节点）。

Keys to Exercises of the Text

Ⅱ. 1. T 2. F 3. F 4. F

Ⅲ. 1. B 2. C 3. A 4. B 5. C

Ⅳ.

1. 4G is the fourth generation of wireless communications currently being developed for high speed broadband mobile capabilities.

2. The 4G technology is being developed to meet QoS (Quality of Service) and rate requirements that involve prioritization of network traffic to ensure good quality of services.

3. The trend is to provide wireless internet access to every portable device that could supply and incorporate the 4G embedded modules.

4. 4G is characterized by higher speed of data transfer and improved quality of sound.

第 11 单元　智 能 手 机

你可能听到过"智能手机"这个术语很多次了。但如果你曾经想要确切地知道智能手机究竟是什么，好吧，其实不只你一个人曾经这样想过。智能手机和一个普通手机相比到底有什么不同，又是什么使它这么智能呢？

简而言之，智能手机是这样的一个设备，它能让您拨打电话，并且还增加了其他一些以前只能在掌上计算机或计算机里才能发现的功能，如发送和接收电子邮件、编辑 Office 文档。

但是，要真正理解什么是智能手机或者什么不是智能机，我们首先应该开始一堂历史课。起初，手机和掌上计算机（PDA）都问世了。手机除了用来拨打电话外，其他的功能不多，而 PDA，如 Palm 公司的 Pilot，扮演着便携的个人事物组织者的身份。一个 PDA 可以用来存储您的联系人信息和待办事宜清单，并且可以与您的计算机进行同步。

PDA 获得了无线连接后能够发送和接收电子邮件。在当时，手机短信功能也出现了。然后，在 PDA 添加了更多手机的功能，而手机上添加了更多类似 PDA（甚至计算机）的功能，结果就是智能手机产生了。

与许多传统手机不同，智能手机允许个人用户安装、配置和运行他们所选择的应用程序（如图 11-1 所示）。智能手机具有了可以使设备遵从您特定的特有做事方式的能力。传统的标准手机软件只为重新配置手机提供有限的选择，使您不得不去适应它的设置方式。在一个标准的手机里，不管你喜欢不喜欢内置的日历应用程序，除了做一些小调整之外，你不得不接受它。如果这是一个智能手机，你可以安装任何你喜欢的和兼容的日历应用程序。例如，智能手机将使您能够做更多的事情。它可以让你创建和编辑微软文件或至少查看文件。它还可以让您下载 APP 应用等。

智能手机通过使用类似于计算机使用的操作系统来运行，一些比较流行的操作系统有安卓、视窗、塞班和黑莓操作系统。

安卓操作系统在 2008 年发布，它被认为是一个谷歌支持的开放源程序的平台。它的功能包括了日历、地图和一个全功能的网页浏览器。苹果公司的 IPhone 手机是非常受欢迎的智能手机。它于 2007 年第一次向全世界发布，而且售价高达 499 美元。它是第一个使用大尺寸触摸屏的手机，其目的是为方便用手指直接输入。

由于手机和掌上计算机是今天最常见的手持设备，智能手机通常既可以说是一个增添了掌上计算机功能的手机，又可以说是一个增添了电话功能的掌上计算机。下面是智能手机可以做的一些事情：

- 呼出和接收呼叫——一些智能手机还具有连接 WiFi 的能力；
- 个人信息管理（PIM）能力包括便条，日历和待办事宜清单；
- 与笔记本式计算机或台式计算机通信；
- 与应用程序进行数据同步，如 Microsoft Outlook 邮件程序和苹果的 iCal 日历程序；
- 电子邮件；
- 即时消息；
- 应用程序，如文字处理程序或者视频游戏；
- 以一些标准格式播放音频和视频文件。

现在多数智能机包含一些软件，即使最基本的机型也包含地址簿和某种形式的联系人管理器。

Keys to Exercises of the Text

Ⅱ. 1. b 2. a 3. d 4. c 5. g 6. e 7. h 8. f

Ⅲ. Demonstrated；prototype；as well as；referred；make；able apps；through；

Ⅳ.

1. 1999 年，日本的 NTT DoCoMo 公司发布了第一款能够达到一个国家内广为使用的智能手机。

2. 塞班是 20 年代中期到年代末欧洲最为流行的智能手机操作系统。

3. 在 2013 年，有报道认为，随着预期制造成本的降低，可折叠的有机液晶屏智能手机将在十年后面世。

4. iPhone 引进的很多设计理念被现代智能机平台所采用。

阅读材料　　　　　　　　　**电子邮件与网上聊天**

如果你上网，你可能使用基于互联网的通信方式与家人、朋友或同事联系。互联网提供了很多联系方法，从给朋友发个即时消息、给同事寄个电子邮件，到打电话、视频会议。

互联网通信有很多有优点。如果你（或你的雇主）已经为相应的互联网应用账户付费，你就可以通过发送即时消息或用网络 IP 语音电话而省去打本地电话的费用，当然，没有一种技术是无缺点的，网络通信也有很多缺点，如病毒、隐私泄露和垃圾邮件等。

与所有技术（尤其是捆绑于互联网的技术）一样，在线通信的方法是在不断发展的。

这里我们谈一下几种最主要的互联网通信方式。

1. E-mail

E-mail 是电子邮件的缩写，是基于通信网的信息传输。许多大型机、小型机及计算机网络都有电子邮件系统。有些电子邮件系统是局限于单个计算机系统或网络中的，但有些可通过网关送到连接到其他计算机系统，可以把电子邮件送到世界的任何一个地方。

用一个电子邮箱软件（如微软的 Microsoft Outlook 或 Eudora），你可以写电子邮件，无论何地何人，只要你知道他的电子邮箱地址就可以给他发电子邮件。所有在线服务和网络服务提供商都提供电子邮件和网关服务，所以你可以与其他计算机系统的用户交换电子邮件。一封电子邮件一般只要几秒钟就可以送到目的地。电子邮件可以对一群人同时发信息或传送文件，这是一种与一群人通信的特别有效的方式。

始终笼罩在电子邮件头顶的一块最大的乌云之一是垃圾邮件。虽然垃圾邮件的定义始终在变，垃圾邮件可以理解为是指送到千家万户的电子宣传品（通常是产品广告的电子邮件），垃圾邮件经常传播特洛伊木马（一种病毒）和病毒。因此，用更新后的杀毒软件对所收发电子邮件进行杀毒是很重要的。

2. 即时消息

即时消息（IM）是互联网中发展最快的通信形式之一。可把 IM 看成是在两人或更多人之间的计算机文字会议，IM 通信服务器可以创建个人聊天室，使你与另一人在网上实时聊天。IM 系统还有提醒功能：当你的网上好友或联系人在线时提醒你，你就可以和某个人开始聊天。

当成百上万个网络用户用 IM 与家人或朋友文字聊天时，IM 在工作中的应用也越来越普遍。当有事要谈时，公司的职员可以与不在一个办公室的经理或同事直接交谈，而不再需要打电话。总之，IM 可以节省职员的时间并减少企业的通信开销。

Keys to Exercises of the Reading

Ⅰ. 1. T　2. T　3. F

Ⅱ.

1. Instant messaging (IM) is a type of online chat which offers real-time text transmission over the Internet.

2. Electronic mail has been most commonly called email or e-mail since around 1993, but various variations of the spelling have been used.

3. Skype, a voice over IP (VoIP) service, was first released in 2003 as a way to make free computer-to-computer calls, or reduced-rate calls from a computer to telephones.

第 12 单元　5G 技术

5G 技术指的是第五代移动通信技术。5G 移动技术已经使我们使用移动电话的方式发生了改变，非常宽的带宽将被使用。用户以前从未体验过如此高价值的技术。现在在移动用户的手机有更多的（手机）技术认知度，5G 技术包含各种先进特征，使得其在不久的将来有巨大的需求。

被纳入到新手机的非常大量的创新技术是惊人的。手持电话中的 5G 技术将提供至少 1000 个月球登月舱才能提供的功能。用户可以将 5G 手机连接到笔记本式计算机来获得宽带接入。5G 技术包括：照相机、MP3 录音、视频播放器、大手机内存、快速拨号、音频播放器和更多您没有想到的功能。对于孩子们来说，摇滚迷蓝牙技术和微微网已经在市场上成形。

5G 技术能够带来什么

5G 技术将在移动产品市场上引发一场革命。通过 5G 技术您可以在全球范围内使用手机，而且这个技术将在中国移动市场引起很大轰动，一个熟练的用户能够进入一个德国的手机并把它当本地手机使用。随着像 PDA 一样的移动电话的出现，现在您的整个办公室业务都在您的指尖下或手机里。5G 技术具有超凡的数据处理能力和不受限制的电话量，并能够使用最先进的移动操作系统进行无限的数据传播。因为能够操纵最好的技术和提供给用户非常有趣的手机，5G 技术将有一个非常光明的未来。5G 技术可能在接下来的日子里接管世界的移动通信市场。5G 技术具有超凡的能力来支持软件和咨询。在 5G 技术中使用的路由器和转换技术将提供很高的连通性。5G 技术分配互联网的接入到建筑内的各个节点，并能够部署有线或者无线网络连接联合体。5G 技术现在的趋势将有一个多彩的未来。

5G 技术特征

- 5G 技术为电话狂用户和双向大带宽定型提供高分辨率。
- 先进的计费界面使 5G 技术更高效和更具吸引力。
- 5G 技术还将提供给订户监督工具以便快速反应。
- 5G 技术提供基于政策层面的高质量的服务来避免差错。
- 5G 技术提供的千兆字节的数据传播能够支持 65000 个连接。
- 5G 的技术提供了无与伦比的一致性转运级网关。
- 5G 技术的数据流统计更准确。
- 用户可以通过 5G 技术提供的远程管理工具来获得更好和更快的解决方案。
- 远程诊断也是 5G 技术的一个重大特征。
- 5G 技术提供达到 25 Mbit/s 的连接速度。
- 5G 技术也提供虚拟个人网络。
- 5G 技术的商业展望中将没有交付服务这一项。
- 上传和下载的速度将达到顶峰。
- 5G 技术将提供世界上可获得的强化的连接。

5G 技术将使普通计算机和笔记本式计算机在竞争中生存很艰难，它们的市场价值将受到影响，所以一个新的 5G 技术革命将要发生。从 1G、2G、3G、4G 到 5G，电信技术已经进步了很多。即将到来的 5G 技术与以往技术相比，将以能够承受的价格和更可靠的方式出现，并在未来达到顶峰。

Keys to Exercises of the Text

Ⅱ. 1. c 2. d 3. g 4. h 5. f 6. a 7. b 8. e

Ⅲ.

1. 移动通信技术的代一般情况下指的是不向后兼容的蜂窝技术标准，这些标准符合 ITU-R 公布的要求，比如 3G 的 IMT-2000 和 4G 的 IMT-Advanced。

2. 5G 技术为电话狂用户提供高的分辨率和双向大带宽定型。

3. 5G 技术先进的计费接口使得它更具吸引力也更高效。

4. 5G 技术能够提供 G 字节的数据传输，这使得它能够支持达 65,000 个连接。

第 3 章　计算机技术

第 13 单元　计算机系统

计算机系统由许多不同的子元器件组成，这些元器件组合起来可进行计算或完成复杂的任务。计算机系统可以管理工资报表,控制汽车发动机,为飞机导航,让用户玩电子游戏,甚至进行账目结算。

计算机设备的大小、成本及功能差异很大,主要取决于它要完成什么工作。本单元主要研究个人计算机系统——适合小型企业和家庭使用的计算机系统。

主机

主机装有计算机的主板,主板上装有计算机存储器(常指随机存储器)和中央处理器。随机存储器能存储程序和数据。随机存储器越大,就能运行越复杂的程序,处理越多的数据。RAM 的容量用"吉字节"来衡量(1GB = 10 亿字符)。目前个人计算机系统的容量一般是 500GB。

软驱和硬驱是用来永久保存程序或数据的存储设备。软驱支持可移动的媒体盘即软盘,用户可以取出软盘放在其他计算机中使用。硬盘是不可移动的媒体盘,因为它安装在计算机主机之中。软驱能容纳 1.44MB 的数据;而固定磁盘(硬盘)能支持高达 500GB(相当于 512000MB 或 5000 亿字符)。

CPU 是实际运行程序和进行数据处理的设备。它像汽车的发动机一样,做所有的工作。电源(Power Supply Unit)也安装在主机内,给存储器、CPU 和计算机的其他设备供电。

键盘

用户使用键盘将命令和数据输入计算机系统。进行支票账目核算或远程拨号,都是用键盘输入命令的例子。

显示器

显示器是计算机向用户显示信息的设备。或显示文本信息,或显示图形信息。显示器有多种型号,有 20 英寸、22 英寸、24 英寸等。显示器越大,价格越贵,屏幕显示的影像也就越大。

显示器有许多重要指标:屏幕分辨率和刷新率。屏幕分辨率指的是在 x 和 y 轴上(1400×900 或 1920×1080)的一些小点。刷新率具体为每秒钟刷新荧光屏图像的次数(60Hz,即每秒 60 次)。在这可以了解显示器的很多内容。像 1400×900 的高荧屏分辨率须选用 21 英寸的显示器(否则如果选用 14 英寸的显示器则看起来太小),当然还需要高达 72Hz 的刷新率,以免荧屏上图像闪烁。

鼠标

鼠标是一个下面嵌有轨迹球的设备,上有一组选择按钮。手握鼠标在鼠标垫上移动时,鼠标的位置就显现在屏幕上。随着鼠标的移动,屏幕上显示出它的位移。按钮用于进行屏幕选择。

通过将使用键盘输入命令改为用鼠标单击屏幕上显示的按钮或其他对象,鼠标可以大大减少用户的输入操作。

打印机

用户用打印机将计算机处理过的数据资料或屏幕信息复制在纸上。有彩色打印机和黑白打印机。彩色打印机比黑白打印机的速度慢且价格昂贵。此外,不同的打印机所使用的技术也不相同。

调制解调器

用户用调制解调器将自己的计算机系统与另一计算机系统相连。调制解调器要接在电话线上,通过电话向另一台计算机拨号联系。

调制解调器转换了计算机信号,因此可以通过电信公司的电话线路传递信号。调制解调器有两种,一种是安装在主机内部的内置设备,另外一种是通过线缆同主机相连的外置设备。

Keys to Exercises of the Text

Ⅰ.

1. A typical personal computer system consists of base unit, keyboard, monitor, mouse, printer and modem.

2. The CPU is the device that actually runs all the programs and processes the data. It's like the motor of the car, it does all the work and makes things happen.

3. Screen resolution refers to the number of dots in the X and Y co-ordinates (640×480, or 800×600).

4. The modem.

Ⅱ.

1. 主机　　　　　　2. RAM　　　　　　3. 中央处理器
4. floppy drive　　　 5. 屏幕分辨率　　　 6. refresh rate

Ⅲ.

inexpensive, individual, type, browse, addition, keyboard, monitor, storage

Ⅳ.

1. 用户接口提供人机通信和交互的途径。
2. 虚拟存储器是用于暂存数据和在需要时和 RAM 交换数据的硬盘空间。
3. 一个小型机能够支持从 4 至约 200 个用户，同时是一个多处理系统。

阅读材料　　　　　　基本输入输出系统

BIOS 是一种特殊的软件，是主要硬件与操作系统的接口软件。通常存储在主板的闪存芯片中，但有时也可能是存储在一种 ROM 芯片中。

BIOS 软件有很多不同的作用，但它最重要的作用是载入操作系统。当你开机时，微处理器开始执行第一条指令，它必须从某个地方得到这条指令（就是储存在闪存中的 BIOS）。微处理器不能从操作系统得到指令，因为操作系统存储在硬盘上，没有指令 CPU 不能从硬盘上载入操作系统，BIOS 提供了这些指令。BIOS 还提供了一些其他指令（程序），如：

- 电源自检（POST）用来检查系统中各个硬件是否工作正常。
- 激活（调用）计算机中其他不同卡上 BIOS 芯片中的程序，如 SCSI（接口卡）和图像处理卡本身自带 BIOS 芯片。
- 提供操作系统与不同硬件接口的低级处理程序，就是这些低级程序调用各种设备。这些程序管理键盘、显示屏、串行和并行接口等设备，尤其当计算机启动时。

BIOS 所做的第一件事是检查储存在一个微小（64 字节）RAM（位于一个 CMOS 芯片中）中的信息。CMOS 给出了你系统的详细信息，当系统改变时可以改变 CMOS 设置，BIOS 要根据这些信息调整或补充它的默认程序。

Keys to Exercises of the Reading

Ⅰ.

1. T　　2. F　　3. F

Ⅱ.

BIOS 所做的第一件事是检查储存在一个微小（64 字节）RAM（位于一个 CMOS 芯片中）中的信息。CMOS 给出了你系统的详细信息，当系统改变时可以改变 CMOS 设置，BIOS 要根据这些信息调整或补充它的默认程序。

第14单元　计算机操作系统

当前台式计算机和小型计算机通常有一个微处理器作为它的中央处理器，工作时，微处理器执行一系列的指令，这些指令称为软件。操作系统就是一种软件。

操作系统为计算机中运行的应用程序提供一系列的服务，还为计算机提供基本的用户接口。

操作系统的目的是组织和控制硬件和软件，使设备的工作方式是可调节和预先设定的。

所有的台式机都有操作系统，最通用的是微软开发的 Windows 系列的操作系统、苹果公司开发的 Macintosh 操作系统和 UNIX 系列操作系统。对一个台式机用户来说，这就意味着你可以安装新的安全更新、系统补丁、安装新的应用程序甚至是新的操作系统，当你想对这些设置进行改变时，完全不需要换一台新的计算机。

一个操作系统至少要做这样两件事。

一是管理系统的硬件和软件资源。因为各种程序和输入方法为各自的目的争抢 CPU、内存、储存设备和输入/输出口，这时操作系统起到一个好的管理作用，它保证各个应用程序得到必要的资源，并与其他应用程序协调工作。操作系统面向所有的用户和应用程序，使系统的有限资源得到最充分的应用。

二是操作系统为应用程序提供了一种稳定、一致的接口（界面），使得处理硬件的应用程序并不需要知道具体的硬件电路。这尤其对不同类型的计算机或硬件不断变化的计算机来说，采用操作系统就显得更加重要。一个一致的应用程序界面（API）使一个软件开发者在一台计算机上编写的应用程序可以在其他任何同类的计算机上运行，即使两台计算机的内存或储存量不同也没关系。

Keys to Exercises of the Text

Ⅱ. 1. T　2. F　3. F　4. T　5. T

Ⅲ. 1. desktop computer　　2. laptop computer　　3. software resources
　　4. UNIX 系列操作系统　5. 系统补丁　　　　　6. 应用程序

Ⅳ.

1. 为了定义操作系统控制进程和资源所需要的控制结构，操作系统在初始化的时候必须拥有对设备数据的控制权。
2. 操作系统的主要任务是控制进程的执行。
3. 典型的 UNIX 系统采用两种运行态来表示进程是运行在用户模式还是内核模式。

第15单元　编程基础

C 通常被看成是一种结构化语言。从学术观点来看，C 只是模块-结构化语言中的一个非正式的部分。模块-结构化语言的主要特征是代码与数据的分离化。这意味着这种语言可以把完成一个指定任务所需要的信息和指令与程序的其他部分分离开，并隐藏起来。

函数是 C 语言的组成模块，在函数中可以进行各种编程。函数可以使编程员在一个程序中分别定义和编码指定的任务。在调试过一个只用局部变量的函数以后，这个函数可以在各种情况下正常运用，不会对程序的其他部分产生副作用。所有在这个函数中声明的变量只能在这个函数中应用。

在 C 语言中，用代码块还可以创建程序结构。一个代码块是一组逻辑上相联系的程序语句，它们可以被看成是一个块。你可以在左右两个花括号之间插入几行代码创建一个代码块。在 C 语言中，每个语句可以只有一个语句，也可以是一个语句块。利用代码块可以创建合乎逻辑的容易理解的程序。

C 语言是一种编程语言。不像其他高级计算机语言，C 语言对程序员能用它做什么没有什么限制。在绝大多数情况下，编程者可以使用 C 语言来替代汇编语言。事实上，发明 C 语言的原因之一就是为了提供一种替代汇编语言编程的方法。

Keys to Exercises of the Text

Ⅱ． 1．block-structured language　　2．a programmer's language　　3．assembly language
　　4．代码块　　　　　　　　　　5．花括号　　　　　　　　　　6．副作用

Ⅲ．input, data, output, mode, digital, microcomputers, equal, speed

Ⅳ．
一种编程语言包括允许人们与计算机交流的所有的符号、角色和使用规则。一些编程语言是为有特殊用途的机器而创建的，而其他的编程语言则服务于更加灵活的通用机，这些机器适用于许多类型的应用程序。

阅读材料　　　　　　　　　　MATLAB 语言

MATLAB 是一种数字化计算环境和可编程的语言。它集成了计算、可视化和编程在一个极易使用的环境中，问题和解答都以人们熟悉的数学符号表示。

MATLAB 是一个交互系统，其基本数据元素是阵列，用它可以很方便地计算很多问题，尤其是含矩阵和矢量公式的问题。另外它还有一个附加 Simulink 软件包，为动态系统和嵌入式系统提供了多领域的图形仿真和基于模型的设计（方法）。

MATLAB 用起来很方便，这里举两个例子。

1．启动和退出

当启动 MATLAB 时，桌面上显示一个含有管理文件，变量以及和 MATLAB 相关应用程序的工具的窗口（图形化用户界面）。

图 15-1 给出常用的窗口，你可以根据你的需要调整工具和文档等窗口。要退出 MAT-LAB，在菜单中选择文件 > 退出 MATLAB，或在命令窗口中直接输入 quit。

2．画图

MATLAB 提供了很多方法显示数据图像，还有交互工具可以用来处理图像，使图像可以反映出关于数据的最多信息。

例如，下面的语句建立了一个变量 x（阵列），取值范围为 –1 ~ 1，间隔为 0.1，第二句语句求出 x（阵列）的每一个值的三次方，并储存为 y 阵列：

x = -1:. 1 :1;　　%定义 x 阵列
y = x. ^3;　　%求出 x（阵列）的每一个值的三次方，并储存为 y 阵列
plot (x, y);　　%画 y – x 图像

一个简单的曲线（图）很好地显示了 x 作为自变量，y 作为函数的关系（见图 15-2）。

你还可以在此图上加注释和打印此图，或输出图片的标准格式用以在网页浏览器或其他媒体上显示此图。

Keys to Exercises of the Reading

1. 选择使用这个工具的原因是，它为开发分析引擎提供易于使用的编辑器和工具。
2. 要以交互模式运行，在提示符中输入每个命令。
3. 这些应用程序既包含交互功能，也包含业务逻辑。
4. 当然，您可以根据自己的喜好来定制页面。
5. 您可以利用这些标签，从命令行操作您的后台工作。

第16单元　多媒体技术

1. 什么是多媒体

动画、声音、图像、文本、视频和照片等都是媒体元素。

多媒体是指结合两种或多种媒体类型，从而生动地表达一种思想的作品，作品通常有声音和视频的支持。典型的多媒体产品使用计算机开发和控制。

多媒体并不是新词，该词已被使用了几十年，用来描述伴随有磁带录音或者同步磁带的幻灯片报告。在商业介绍中，幻灯片和解说相结合的形式已经十分流行和成功。

在20世纪70年代，幻灯片放映方式被引入到计算机中。计算机可控制多个投影仪，计算机技术配合这些投影仪产生一种快节奏的切换方式和生动的效果。磁带声道中含有信号，这些信号使投影仪按事先设置的方法放映。

到20世纪80年代，计算机已可以切割图像元素，并能把图像元素"粘贴"到另一个文件中。从那以后，软件和硬件开发商们开始争相把各种形式的媒体集成到个人计算机中。

2. 多媒体资源

生动有效的多媒体效果依赖于对各种各样的被称为资产或者资源的材料的充分使用。多媒体资源有各种各样的形式。

（1）文本

与纸质文本相比，屏幕上的文本有三个主要的优点：能自然地刷新屏幕、具有反应能力，并可以产生特殊的伴随效果。

（2）声音

交互的声音可以给多媒体系统增加一个特别真实的元素。以前，基于声音的领域（如音乐、语言学、语言等）实际上被忽略了。如果能更快地实现声音交互和音频质量的提升，那么人和计算机之间的交流将变得更加生动。

（3）图片

当要使用图片时，设计者必须选定最经济有效的方法去制作图片。在多媒体应用的规划阶段，设计者必须选择技术和软件最适当的组合，从而再现和产生新的图像。

（4）视频

手提式的视频录像机可以很方便地拍摄实时视频图像。通过两种不同的方法可以把视频与多媒体应用结合起来。其一是通过控制卡把视频源连接到计算机上，这种技术称为交互视频。更完整的方法是把视频从模拟信号转换成可以被计算机处理的数字格式。

3. 多媒体的应用

多媒体有各种各样的分类方法，可以按市场（如家庭、商业、政府和学校）、用户（如儿童、成年人、教师和学生），或者门类（如教育、娱乐和参考资料等）分类。这里我们把它们归为7大应用领域，并逐一详细阐述。

（1）参考资料

CD参考资料有很多，如百科全书、户口普查资料、黄页、地图集和街道目录等。在大多数情况下它们是参考书的电子版本。开发者要做的是使用户可以很方便地查到想要的信息，并有效地利用其他媒体元素，如声音、视频和动画。

（2）教育

教育者的目标是促进学习——帮助学生获得整体的知识、获取特定技能、成功地立足于社会。教育者面临的最大挑战就是学生的多样性，尤其是他们的学习方法是各种各样的。有些学生借助于联想会学得比较好，有些学生则借助于实验能学得更好，有些喜欢多看，有些喜欢多听。

多媒体可以适应不同的学习模式，可以用非线性化的方式展现材料。它是激发式的，具有高度的交互性，还可以提供反馈和评价的技能。

（3）培训

每个公司需要在从人员政策到设备维护等很宽范围内培训各自的雇员。

受培训者可以用多媒体完成仿真的工作任务，从而进行高级培训，而不必接触实际设备。声音和视频的结合可使诸如飞行和驾驶领域的技术培训变得非常生动。美国国家航空宇航局就广泛使用这种多媒体技术对宇航员进行飞行控制训练。

（4）娱乐

很难划分多媒体中教育与娱乐的界线，因此就有了新词"寓教于乐"，多媒体可以使学习变得非常有趣。

但多媒体也有纯娱乐的一面。声音和图像所具有的效果在任何一张多媒体 CD 上都可以实现。AIATSIS 是一本澳大利亚土著居民的百科全书，具有 2000 多项条目——1000 张图片、230 个声音片段和 50 段视频录像，其内容包罗万象：从艺术到健康，从技术到法律。

（5）商业

商业有与外部世界交换信息的需求。多媒体为商品介绍、市场和销售提供了很广的选择范围。多媒体可用于贸易展示，或生成电子目录。利用多媒体可以增强新产品的上市推广，这些产品可以比使用印刷品提供更详细和更具吸引力的产品信息。

（6）演讲

每天在商界中有成千上万个多媒体讲演。公司的首席执行官们在股东会议上作年度报告，销售代表向潜在的顾客兜售系列产品，会议的演讲人在给听众介绍工业趋势。从电子幻灯放映到互动的视频展示，多媒体使演讲更丰富多彩。

多媒体给演讲人提供了可以吸引听众、集中注意力、强化关键概念和使演讲变得生动的工具。

（7）互动游戏

多媒体意味着互动，许多互动娱乐则意味着游戏。在利用多媒体技术方面，游戏开发商总是走在前面，直到现在仍然提供着最新的和交互的多媒体技术应用软件。

为了吸引玩家、留住玩家、迷住玩家、挑战玩家，多媒体技术提供了娱乐必需的快速的动作、逼真的色彩、三维动画和完美的声音效果，它还提供了奖励、赞誉和成就感等娱乐常有的部分。

许多游戏已从体力（手、眼配合）转移到脑力（解谜、战胜恶魔、与对手斗智）。

另一方面，多媒体也可以应用于业余爱好和运动，如让用户体验世界上最好的高尔夫课程或在三维城市的高空作模拟飞行。

Keys to Exercises of the Text

I

1. audio　　　　　　2. multimedia　　　　3. find the desired information
4. video　　　　　　5. slide　　　　　　　6. provide feedback
7. entertainment　　 8. 3D animations

II

1. 产生一个有序事件　2. 多媒体资源　　　　3. 便携录像机
4. 数字视频　　　　　5. 善于联想　　　　　6. 有纯粹娱乐的一面
7. 最经济有效的方法　8. 技术和软件最佳的结合方式

Ⅲ.
电路仿真的 4 个步骤

EWB 仿真软件中的仿真器与其他通用仿真器类似，有 4 个主要的阶段：输入、组织、分析和输出。

在输入阶段，当你设计完电路、分配好参数并选择一种分析方法后仿真器开始读取电路信息。

在组织阶段，仿真器构建并检查完整描述电路的一组数据结构。

分析阶段分析输入阶段规定好的电路。这个阶段占用了 CPU 大部分的执行时间，并且是电路仿真的核心。分析阶段把指定的电路用公式表示并求解，最后提供直接输出或者后续处理的数据。

在输出阶段，你会看到仿真结果。在运行分析时出现的菜单上，或者在选择分析/显示图形时，你都可以在诸如示波器等设备上查看到结果。

第 17 单元　中国在超级运算上的进步

中国国防科技大学研制的中国"天河一号"超级计算机已经名列世界前 500 超级计算机榜的顶级位置。这是中国科学技术快速发展的一个重要标志。

天津国家超级计算机中心的天河 –1A 系统可以执行每秒 2.57 万亿次的计算。

星期二，美国和欧洲的研究人员正式公布了世界前 500 强超级计算机的名单，中国占了 41 台。在前 10 计算机中，中国有 2 个系统入围，这十台中有 7 台达到甚至超过每秒一千万亿次的运算能力。美国有 275 台计算机上榜，法国、德国和日本各占 26 台，英国 24 台，俄罗斯 11 台。

考虑到超级计算机领域的激烈竞争，这是中国的一次巨大成功。超级计算机是一个国家在很多领域进行科学研究的基础工具，这些领域包括：地理、气象和石油勘探以及航空、汽车、化工等。

这份名单是由德国曼海姆大学的汉斯摩尔，国家能源研究科学计算中心（NERSC）/劳伦斯伯克利国家实验室的施特罗迈尔和赫斯特西蒙，诺克斯维尔的田纳西大学的杰克唐加拉编写的。

在 20 世纪 90 年代初，中国几乎所有的计算机都是进口的，科学研究机构只能在外国专家的指导和管理下使用计算机。

近年来，在"863 计划"的支持下，中国研究人员努力研发中国自己的超级计算机。"863 计划"是一个政府出资的国家级高新技术研发倡议。

这些"中国制造"的超级计算机不仅打破了外国的封锁，也缩小了中国与发达国家的差距。他们甚至使中国在一些技术上处于世界领先地位。

在一定程度上，超级计算机反映了一个国家对计算能力的需求，以及一个国家的基本研究以及高技术产业的发展水平。法国原子能委员会的一位官员指出，"天河一号"的计算速度显示了中国经济的竞争力已经增强。

然而，需要记住并引起注意的是，在整体计算能力方面，尤其是核心电子和高端芯片的发展方面，中国仍落后于像美国这样的发达国家。

美国和日本正在研究比"天河一号"性能更好的超级计算机，而且在未来的发展中，各国在超级计算机的研究中会轮流占据领先地位。要超级计算领域赶超其他国家，中国还面临着艰巨的任务。

然而，"天河一号"已经使"中国制造"在超级计算机领域占领一席之地，也传递出中国科学技术正在快速发展的信号。

Keys to Exercises of the Text

Ⅱ. 1. c 2. h 3. b 4. g 5. d 6. a 7. f 8. e

Ⅲ. ranked; until; overtook; announced; simulation; that; ran; achieve

Ⅳ.

1. 这个超级计算机运行时消耗4.04MW的电力,103个计算机柜占地17000平方英尺。

2. 新闻媒体和评论员们讨论美国怎样被从保持了6年之久的计算机系统500强名单的首位位置上被挤了下来。

3. 它号称具有2.507千万亿次/秒的运算速度,打破了Cray XT5 Jaguar保持的纪录。

阅读材料　　　　　　　　　　比尔·盖茨

　　威廉〔比尔〕·H. 盖茨是微软公司主席、首席软件设计师,也是全世界个人及商务计算机领域软件制作、服务、互联网技术的领军人物。微软公司现有4万多名员工分布在世界上60多个国家。到2001年6月底截止的本财政年度总收入为253亿美元。

　　盖茨1955年10月28日出生,与他的两个姐妹一起在西雅图长大。盖茨小学就读于一所公立学校,中学转到雷克赛德学校,这是一所私立中学。在那里,盖茨对软件发生了兴趣,13岁时便开始为计算机编制程序。

　　1973年,盖茨就读于哈佛大学,和史蒂夫·巴尔墨住在同一栋宿舍楼里,后者现在是微软公司的首席执行官。在哈佛读书期间,盖茨为第一台微机——MITS Altair编制了一套BASIC语言。

　　大学三年级时,盖茨从哈佛退学,专心于创建微软公司。早在1975年,盖茨便与童年时的朋友保罗·艾伦一起筹建微软公司。他们坚信计算机将是每张办公桌上和每个家庭里非常有用的工具。带着这一信念,他们着手开发个人计算机软件。盖茨对个人计算机的远见卓识是微软公司和软件业成功的关键。

　　1999年,盖茨写了一本名为《未来时速》的书。书中介绍了计算机技术怎样以全新的方式解决商务问题。此书有25种语言的版本,在60多个国家均有售。《未来时速》得到评论界的广泛赞誉,被《纽约时报》《今日美国》《华尔街日报》和"亚马孙"网站列入各自的畅销书目录。盖茨早些时候(1995)出版的另一本书《未来之路》曾连续七周高居《纽约时报》畅销书排行榜榜首。

　　除了对计算机和软件情有独钟外,盖茨对生物技术也有兴趣。他是数家生物技术公司的投资人,还创立了致力于开发世界上最大的视觉信息资源的考比司公司。此外,盖茨与移动电话的先驱克拉格·麦考共同投资了Teledesic公司,这家公司正在实施一个雄心勃勃的计划:使用数百颗低轨道卫星为全世界的用户提供双向的宽带远程通信服务。

Keys to Exercises of the Reading

1. In the private Lakeside school he found his interest in software, and began programming computers at the age of 13.

2. He developed a version of programming language BASIC for the first microcomputer.

3. With his childhood friend Paul Allen, Gates started to build Microsoft.

4. Gates' anticipation and his vision for personal computing have been central to the success of Microsoft and the software industry.

5. Gates also founded Corbis, which is developing one of the world's largest resources of visual information.

第4章 高级电子与通信技术

第18单元 人工智能

自第二次世界大战以来,计算机科学家们一直试图开发能让计算机行为更像人的技术。所有这些研究内容,包括决策系统、机器人设备(见图18-1)和实现计算机语音的各种方法,通称为人工智能(AI)。

人工智能最终的目标是开发一种能够从概念和实事中进行学习,做出符合常理的决策,并能进行某种计划的计算机系统。也就是说,最终目标是制造一台能够"思维和学习"的计算机。

计算机程序是一组能使计算机对信息进行处理并做出决策的指令。大多数程序相当刻板——它们精确地告诉计算机每一步都执行什么。但人工智能程序并不遵循此规律。它们会寻求捷径,做出选择,探索和试验不同答案,并改进解决方案等。

很多人工智能程序能够将相关事实组织在一起,使计算机有可能把不同信息相互关联,并和某个给定的课题也关联起来。"如果/那么"这种推理规律也被编入程序,让计算机可以取舍、组织和更新信息。按照这个规则,如果某事为真,则另一些确定的事必随之发生,每一步的行动必会引起一些新的可能的行动。

在电子计算机诞生的初期,弈棋程序就已经出现了,但比较刻板,且受编程人员能力的限制。走哪一步棋,怎样应付对方的棋步等具体的指令都被固化在程序内。有时程序中也会纳入一些象棋专家的建议。但这样的程序很难击败象棋高手。通常在开局时计算机程序较强,但随后越下越弱。

因为人工智能的出现,所有这些都改变了。如今开发的计算机弈棋程序能够战胜大多数象棋高手——包括象棋大师。

当然,人工智能不仅仅会下棋。计算机科学家正在为人工智能寻找几十种实际用途。其中包括操控机器人、解答数学题和科学难题、理解语言、分析图像等。

人工智能程序的最大用途也许是对复杂系统的故障寻迹提供专家咨询(即确定故障所在并予以修复),所谓复杂系统包括像内燃机、核潜艇和人体。换句话说,这些人工智能程序能够排查故障、划定故障范围、区分故障类型,并且提供咨询。

人工智能专家系统不只是应用于特定机械的故障寻迹。现在正在开发的人工智能程序还可应用于经济规划、天气预报、石油勘探、计算机设计以及其他很多用途。

人工智能技术也能用于人类语言的分析和合成。目前正在开发的人工智能技术还借助于激光传感器分析图像信息,以提高机器人性能。

大部分人工智能系统都与某种集成技术有关,比如语音合成技术与语音识别的集成。人工智能系统集成的核心思想是使单个软件组件,例如语音合成器,能够与其他成分,诸如常识知识库进行交互操作,以创造更大、更广泛和更强的人工智能系统。集成的主要的方法已被提出,包括消息路由、软件之间相互沟通的通信协议等,它们常运用中间件黑板系统来进行(见图18-2)。

Keys to Exercises of the Text

Ⅰ. 1. F 2. T 3. F 4. T 5. T

Ⅱ. 1. c 2. b 3. a

Ⅲ.

1. Thanks to AI research, all that has changed.

2. A computer program is a set of instructions that enables a computer to process information and solve problems.

3. AI techniques are also being used to analyze human speech and to synthesize speech.

4. Most artificial intelligence systems involve some sort of integrated technologies

Ⅳ.

AI 是一个正在发展的包括许多学科在内的领域，AI 的分支领域包括：知识叙述、学习、定理证明、搜索、解答问题以及规划设计、专家系统、自然语言（文本或语音）理解、计算机视觉、机器人和一些其他方面（如自动编程、AI 教育、游戏等）。AI 是使技术适应于人类的钥匙，它将在下一代自动化系统中扮演更为关键的角色。

第 19 单元　传感器技术

传感器是一种能够将外界的输入量转换成电或光输出信号的装置。

传感器和传感器系统能执行各种各样的检测功能，并能够采集、获取、传递、处理和分配物理系统的状态信息。这些信息可能是化学组成、结构和构造、大型结构、位置等。现代社会极少有产品和设备不用到传感器。

1. 传感器的价值链

传感器技术在各学科间有明显的区别。很少有构件完全有能力在内部实现传感器的功能。

传感器产品的实现需要完成一些任务，这些任务涉及从产品的确定到最终成品以及之后的销售和服务。图 19-1 就是传感器开发的价值链及其主要的应用领域。

2. 无线传感技术

卫生保健、工业自动化、消费品和安全等领域对无线自供电的传感器的应用需求很大并且呈不断增长的趋势。射频识别技术便是一个实例，显示了巨大的应用潜力。无线连接和没有内部电源的传感器预期将在卫生保健、消费品、结构健康监测和能源发掘等领域发挥重要的作用。

图 19-2 是一个传感器现场，其中包含了大量的传感器连接节点。每一个节点都由无线传感器构成，通常不含内部电源。这些传感器与无线收发器相连接，再与下属元器件相联，有可能是一个所谓的接收器。数据的收集由管理设备控制。

3. 生物传感器

这是又一个市场有望在随后几年呈现很强的增长趋势的领域。指纹识别设备和虹膜扫描设备就是实例，它们是由人们不断提高的安全方面的需求而发展起来的。

4. 非插入式传感器和非接触式传感器

越来越多的应用要求使用非接触式传感器。光和声音在这个领域独立或共同发挥着很重要的作用。在此，为了克服光或声音独立使用时的固有局限，超声波和光的组合运用变得越来越重要。

5. 小型化和集成化

大型的生产设备经常要用到传感器，但是在线检测却只能在有限的空间进行。恶劣的环境需要优质的传感器，而其优良性能可以通过小型化和集成化来得到。在光学方面，这意味着新的系统和光纤的使用。在声学方面，新的非接触式声激励装置正在研发。

6. 新型材料

如果想以足够低的成本实现大量完全不同的功能要求，就需要用到新型材料。它可能是一些综合使用了光生成和探测技术的微流控系统。高分子聚合物预期将在这个领域扮演越来越重要的角色，不仅是因为它成本低，还由于它功能特性上具有很高的柔韧性。新近发展的先进的显微镜已经能够看到甚至可以移动单个的原子和分子。这就为人们创造全新的材料和加工工艺提供了机遇。这就是人们通常所说的纳米技术，并且目前已受到了广泛的关注。

能够将这一领域的发展达到工业化程度的公司还不多，很多都只是在进行积极的研究工作，并寄予着很高的期望。

7. 传感器整合和传感器网络

复杂的系统通常是用大量不同类型的传感器进行监控。比如 X 射线就可以在线显示焊接的质量，光可以观测化学组成和宏观动力学，而超声波能提供系统内部构造信息。

传感器技术结合了多个检测信息以获得新的功能。此外，使用大量与网络连接的低成本传感器的系统也变得越来越重要。

Keys to Exercises of the Text

Ⅰ．1．超声波　2．能源开发　3．指纹识别　4．微观的　5．宏观的　6．纳米技术
　　7．卫生保健　8．消费品

Ⅱ．1．T　2．F　3．F　4．T　5．F

Ⅲ．
如果想以足够低的成本实现大量完全不同的功能要求，就需要用到新型材料。它可能是一些综合使用了光生成和探测技术的微流控系统。高分子聚合物预期将在这个领域扮演越来越重要的角色，不仅是因为它成本低，还由于它功能特性上具有很高的柔韧性。新近发展的先进的显微镜已经能够看到甚至可以移动单个的原子和分子。这就为人们创造全新的材料和加工工艺提供了机遇。这就是人们通常所说的纳米技术，并且目前已受到了广泛的关注。

阅读材料　　　　　　　　　遥感技术

遥感是在不与物体直接接触的情况下收集物体或地形特征的有关数据的过程。大多数遥感利用飞机或卫星完成。在此过程中，使用仪器测量由地面发射或反射的电磁辐射（EMR）（如图 19-3 所示）。换句话说，遥感就是检测并测量化远处各种材料的物体所放射的电磁能量。其目的是为了能够根据类型、物质和空间分布对物体进行识别和分类。

遥感装置可分为有源和无源两种。有源系统，像雷达和声呐，对目标发射人为产生的能量，然后将反射的分量记录下来。无源系统，包括照相机，只检查由目标对象自然反射的能量，如反射的太阳光或热红线外辐射。目前，常常使用飞机和地球轨道航天器来携带遥感装置，包括声呐设备。

为了完成遥感的过程，必须采用编译和测量技术对遥感系统捕获和记录的数据进行分析，才能获取有关研究对象的有用信息。这些技术是多种多样的，从传统的视觉判断方式到复杂的计算机处理。所以，遥感系统主要的两个组成部分就是数据采集和数据分析。

目前的遥感技术主要分支有两个。我们把前面首先提到的技术叫作图像遥感，因为它利用图像方面的数据，采用在很大程度上依赖于对图像的生成的分析方法。第二个分支成为数字遥感，因为计算机的发展使它受益而且强调数据的数量，将测量数据抽象为一个集合。在这种情况下，我们并不把图像作为具体的数据来对待，而是把它当作一种便于观察数据的方便途径。

今天我们正通过地球轨道卫星上获得地面观察资料，因为卫星的高度具有宽广的视野，卫星推动传感器运行速度快，所使用的谱带数目多，产生数据数量非常大。

有两种方式可以实现卫星遥感：

- 采用无源传感系统：包括一排小型探测器或传感器阵列，可以检测自地球表面发射

的电磁辐射。
- 采用有源传感系统：系统向一个或多个目标对象发出电磁辐射并测量回波信号的强度。

卫星收集的数据被传送至地面站，在那里，将重组地球表面的图像以获得所需的信息。

卫星遥感的几条优点：
- 对数据的采集可连续进行；
- 有助于获取最新信息（通过编程卫星遥感技术能够定期地重新访问所研究对象或区域）；
- 提供广泛的区域覆盖和良好的光谱分辨率；
- 提供精确的数据信息与分析。

Keys to Exercises of the Reading

Ⅰ. 1. instruments　　2. passive; emitted　　3. transmitted; reconstituted

Ⅱ.
1. 即使你已经知道一个字符串是一个字符数组，你也应该阅读本部分。
2. 当贼使用计算机的时候，计算机就会向备份服务器传输文件。
3. 但直到那时，他还在研究重组蚕丝，并利用它来制造新颖的薄膜和其他材料。

第20单元　物　联　网

物联网又名传感网，是指将各种信息传感设备与互联网结合起来而形成一个巨大网络，这个网络可使所有的物品与网络连接，以便于身份识别和管理。因其具有全面感知、可靠传递和智能处理的特点，它被认为是继计算机、互联网和移动通信网之后的又一次信息产业浪潮。

轻触一下计算机或者手机的按钮，即使在千里之外，你也能了解到某件物品的状况或者某个人的活动情况。发一个短信，你就能打开风扇；如果你的住宅被人非法侵入，你还会收到自动电话报警。这些已不只是好莱坞科幻大片中才有的场景了，物联网的这些场景正逐步逼近我们的生活并变为现实。

这些能够被实现是因为物联网里有一个叫作射频识别（RFID）的存储物体信息的关键技术。RFID 系统由三部分组成（如图 20-1 所示）：天线收发器（通常集成在一个读入器里）和应答器（标签）。天线发出无线电信号激活标签并读出和写入数据到标签。当被激活后，标签发射返回数据到天线。标签传输的数据可能提供识别或定位信息，或者是有关产品标签的信息，如价格、颜色或购买日期等。低频 RFID 系统（30~500kHz）的传输距离较短（一般小于 6 英尺）。高频射频识别系统（850~950MHz 和 2.4~2.5GHz）的传输距离更长（超过 90 英尺）。通常，频率越高，系统的价格则越昂贵。

比如在手机里嵌入 RFID-SIM 卡，手机内的"信息传感设备"就能与移动网络相连，这种手机不仅可以确认使用者的身份信息，还能缴纳水电燃气费、彩票投注、飞机票订购等多种支付服务。

只要将特定物体嵌入射频装置，传感器和其他与互联网连接的设备连接后就能够形成一个庞大的网络系统。在这个网上，即使远在千里之外，人们也能轻松获知和掌控物体的信息（如图 20-2 所示）。

更具体地说，让我们想象一个世界，大量的事物"自动"地围绕着我们，因为它们有：
- 一个名字：具有一个独特代码的标签。
- 存储器：存储那些不能立即从网上获得的任何信息。
- 一种通信方式：移动的和高能效的，如果可能的话。

- 传感器：为了与环境互动。
- 获取或内在的行为：根据主人给出的目标，按特定逻辑行动。

当然，像地球上的其他事物一样，这些都必须以电子信息的形式存在于一个网络当中。

一些专家预测，10年内物联网就可能非常普及并发展成为上万亿规模的高科技市场。然后在个人健康、交通控制、环境保护、工业监测及老年护理等几乎所有领域发挥作用。也有专家表示，只需三到五年时间，物联网就会改变人们的生活方式。

互联网有很大的前途，但这个系统被广泛接受之前，业务、政策和技术上的挑战必须解决。早期应用者需证明，这种新的传感器驱动的商业模式将创造卓越的价值。行业组织和政府监管机构应该研究数据隐私和数据安全的规则，特别是在那些使用接触到消费者的敏感信息时。在技术方面，传感器和执行机构的成本必须下降至某个水平才能实现广泛的使用。网络技术和相应的标准，必须改进到支持数据在传感器、计算机和执行机构间自由流动。收集和分析数据的软件，以及图形显示技术，必须改进提高到这样一种地步——海量的数据可以被人们接收。

Keys to Exercises of the Text

Ⅱ. 1. h 2. g 3. a 4. d 5. c 6. b 7. f 8. e

Ⅲ. physical; embedded; enable; collect; defined; across; creating; integration

Ⅳ.

1. RFID系统由3部分组成：天线、收发器（通常集成在一个读入器里）和应答器（标签）。

2. 只要将特定物体嵌入射频标签，传感器和其他与互联网连接的设备就能够形成一个庞大的网络系统。

3. 使嵌入式装置与有限的CPU、内存和动力源结网的能力，意味着物联网几乎在每个领域都有应用。

第21单元 工业4.0介绍

工业4.0起源于制造强国德国。然后这个概念性的想法已经被很多其他工业国家广泛接受和采用，这些国家包括欧盟内部的成员国，更远到中国、印度和其他亚洲国家。工业4.0这个名字指的是第四次工业革命，前三次工业革命分别通过机械化、电力和IT技术实现了。

第四次工业革命，也因此称为4.0，将通过物联网技术和互联网服务与制造环境的集成而发生。以前的工业革命带来的好处已经既成事实，而我们有机会积极地引导第四次工业革命的发展路径来改变我们的世界。

工业4.0的远景是，将来的工业企业将建立全球网络来连接他们的机器、工厂和库房设施来形成一个信息物理系统，这个系统将通过分享信息来引发动作，从而智能地连接各个部分并进行相互控制。这个信息物理系统将呈现智能工厂、智能机器、智能仓储设施和智能供应链等各种形式。这将带来工业过程的进步，在这个过程中，生产制造作为一个整体，这个改进将通过工程、材料使用、供应链和产品生命周期管理的提升来实现。这也是我们所称的水平价值链，对水平价值链的展望，是工业4.0将在水平价值链的每个阶段进行深度集成，从而使工业生产过程发生巨大的变化。

这个远景的中心是智能工厂，它将使生产方式发生改变，这个改变基于智能机器也基于智能产品。不仅智能工厂等信息物理系统是智能的，被组装的产品也包含嵌入式智能，这样的话它们就能够在生产过程中的任何时间都能被标识和定位。缩微化的无线射频识别标签，使产品变得智能并能够知道它们是什么，什么时候被生产，更为关键的是，它们目前的状况是什么和达到期望的状态还需要的步骤。

这就需要智能产品能够知道它们自己的历史和把它们变成一个完整的产品还需要的工序。这些工业生产制造的知识将嵌入到产品中且允许它们提供生产工艺的替代路线。比如，当智能产品知道了它当前的状态和成为完整产品所需的后续生产过程时，智能产品将能够指挥输送带，这也是它所要跟随的生产线。然后，我们将看看这个工作实际上是怎样运行的。

然而，现在我们需要看看工业 4.0 远景的另一个关键因素，那就是价值链中垂直制造过程的集成。(人们)所持的远景是：嵌入式水平系统与垂直业务过程（销售、物流、财务和垂直业务过程之中的一些其他项）和相关的 IT 系统的集成。它们将使得智能工厂能够对从供应链到服务和产品生命周期管理的整个制造过程进行首尾相连的控制。这个运营技术和信息技术的融合不是没有问题，我们在早些时候讨论工业互联网时就发现了这个问题。然而，在工业 4.0 系统里，这些独立存在的实体将扮演同一个角色。

智能工厂不仅仅是与大公司相关，实际上他们的灵活性使他们更适用于中小企业。比如，对水平制造过程和智能产品的控制，使得我们能够更好地进行决策和进行动态过程控制，在能力和灵活性方面来适应最新的设计变更或者改变生产，来满足顾客在产品设计方面的嗜好。更进一步说，这个动态过程控制使小批量（生产）成为可能，而小批量（生产）还是盈利的，且能够适应个性化订单。这些动态业务和工程过程使得创造价值的新方法和创新的商业模型成为可能。

总的来说，工业 4.0 需要信息物理系统集成在制造和物流过程中，同时在制造过程中引进物联网和互联网服务。这将带来新的方法来创造价值和商业模型，并为下游的中小企业提供服务。

Keys to Exercises of the Text

Ⅱ. 1. f 2. d 3. a 4. b 5. g 6. e 7. h 8. c

Ⅲ.

1. 第一次工业革命通过使用水和蒸汽的力量实现了机械化。
2. 第二次工业革命始于亨利·福特在 1913 年引进的组装线，它使产能大幅提升。
3. 第三次工业革命是在 20 世纪 70 年代计算机导入到生产现场的结果，这引起了自动组装线的大量出现。
4. 工业 4.0 的展望是"计算机－物理生产系统"——在这个系统中布满传感器的智能产品告诉机器他们将被怎样处理。

阅读材料　　《中国制造 2025》和德国工业 4.0 的合作机会

随着中国政府对升级中国大陆工业的计划框架《中国制造 2025》战略的发布，2016 年 3 月采用的"十三五"规划已经开始这个战略，以在下一个五年计划（2016—2020）进行纵深部署。上述行动也唤起了大家对中国工业发展方向的兴趣，一些工业观察者已经把这个战略与德国为提高其工业效率设计的工业 4.0 战略相提并论。

值得一提的是已经有人提出了担心，这两个战略将使得中国和德国工业之间的竞争加剧。尽管如此，两国还是在 2015 年的 7 月签署了加强在智能制造技术开发方面的合作谅解备忘录。

实际上，两个国家的工业方面的相对发展阶段不同，发展相对，再加上不同的战略发展侧重点，相对于竞争而言显示出了更多的合作机会，包括在工业机器人方面的合作。此外，中国和德国工业在全球供应链中的不同位置，在预示着中外合作项目方面的相关选手的更多机会。

本质上来说，德国的工业4.0倡导在生产中采用最新的信息和通信技术来提升工业效率。这个战略的发展以德国强大的机器和设备制造业为基础，其在IT方面的竞争力和嵌入式系统方面的专业技能，以及在自动化工程方面的竞争力使其在制造工业处于很有利的位置，其地位也因此得到加强。

工业4.0的目标是：连接现有的嵌入式IT生产技术与智能过程，来转变和升级工业价值链和商业模型，从而实现智能生产。这将要求德国在加强某些方面的研发，例如深化集成制造系统的研发。这也需要新的工业和技术标准来连接不同公司和装置的系统，同时数据安全系统也应该升级，从而防止系统中数据信息误用和未授权访问。所有这些发展都被期望于加强德国工业的效率和创新能力，同时节约资源和成本。

至于《中国制造2025》，其核心是创新和质量，引导中国工业从低附加值活动向中等或高附加值制造运营活动转变，而不是一味追求产能的扩张。淘汰低效和过时的产能，帮助企业开展更多的自有设计和自有品牌业务，也是这个战略的目标。这些目标将通过以下行动来推进：建立制造创新中心，加强知识产权保护，建立新的工业标准，促进重点和战略部门的发展。

Keys to Exercises of the Reading

Ⅰ. 1. F 2. T 3. T 4. T 5. T

Ⅱ.

1. Industry 4.0 is currently more of a vision than a reality, but it is one with potentially far reaching consequences.

2. Programmable logic will become increasingly important since it will be impossible to anticipate all the environmental changes to which control systems will need to dynamically respond.

3. Whether revolution or evolution, industrial production is about to become more efficient.

第22单元　3D打印技术

增材制造（AM）是描述这种3D物体生成技术的一个恰当名字：通过一层一层地添加材料来生成物体，这些材料可以是塑料、金属、水泥甚至有一天会使用人体的组织。

AM技术的共性是使用计算机、3D造型软件（计算机辅助设计或CAD）、机器设备和层生成材料。在CAD框架生成后，AM设备从CAD文件中读出数据，然后，生成层或者在基材上添加生成连续的液体层、粉末层、片状材料层或者其他材料层，以一层上面又一层的形式来制造3D物体。

AM这个术语包含很多子集，像3D打印、快速造型（RP）、直接数字制造（DDM）、层叠制造和增材制造等。

AM技术的应用是无限的。早期AM的应用侧重于快速原型形式中的预生产可视化模型。最近以来，AM正用于制造飞机，牙科修复物，医疗植入物，汽车，甚至时尚产品中的最终用途产品。

虽然一层一层的材料增加方法是简单的，但是AM技术的许多应用具有各不相同的复杂程度以满足多种需要，包括：设计中的可视化工具，为消费者和专业人员创建高度定制产品的手段，如工业模具，生产小批量零件，将来某一天……生产人体器官。

在增材制造（AM）技术的发明地——麻省理工学院，有大量的项目来支持一系列的前瞻性的应用，从多结构混凝土到可以制造机器的机器；而（轮廓工艺）支持制造人们可以生活和工作在其中的建筑结构。

一些人将AM想象成基础性的减成制造（如采用钻削来去除材料）和相对更小程度的

成形制造（如锻造）方式的补充。无论如何，AM 可以提供消费者和专业人员如下的可能性：创建、定制和/或修理产品，并且在该过程中重新定义当前的生产技术。

无论是简单还是复杂，AM 确实令人惊叹，以一层一层添加的方式对这个令人惊叹的工艺进行了最好呈现，无论是采用塑料、金属、混凝土或者有一天……人体组织。

这里列举一些增材制造的例子。

SLA（立体光固化成型法）

这是一个非常高端的利用激光来一层一层地固化光聚合物树脂（当曝光时改变性质的聚合物）层的技术。

构建发生在树脂池中。激光束被引导到树脂池中树脂上，跟踪具有特定横截面图案的模型的特定层并且固化它。在构建周期内，用于在其上进行构建的平台将重新定位，每次降低一个单层厚度的高度。该过程不断重复，直到模型构建完成并呈现出迷人的外观。可能需要专门的材料来支撑某些模型的特定特征。模型可以进行加工并作为注塑、热成型或其他铸造工艺的样品。

FDM（熔融沉积建模）

这种工艺定位于使用热塑性塑料（当施加热量时变成液体并在冷却时固化成固体的聚合物），通过转位喷嘴注射到平台上。喷嘴跟踪并生成具有特定剖面形状的热塑性塑料的特定层并进行固化，然后进行下一层。该过程不断重复，直到模型构建完成并呈现出迷人的外观。可能需要专门的材料来支撑某些模型的特征。与 SLA 类似，模型可以进行机械加工或用作样品。

MJM（多重喷射造模）

多重喷射造模与喷墨打印机类似，有一个能够来回穿梭移动的（3D x，y，z）的打印头，通过数百个小喷嘴的喷射来形成热塑性塑料层，一层一层地生成物体形状。

3DP（三维打印）

这涉及在填充有淀粉或石膏基粉末材料的容器中构建模型。喷墨打印机头快速地移动来施加少量黏结剂以形成一个特定层。当一层黏结剂施加完成，马上在这一层黏结剂上撒一层粉末，然后施加另一层黏结剂。该过程不断重复，直到模型完成。由于模型是由散粉支持，过程中不需要支撑。此外，这是唯一的一个可以是彩色的模型构建过程。

SLS（选择性激光烧结）

有点像 SLA 技术，选择性激光烧结（SLS）利用高功率激光来熔化塑料、金属、陶瓷或玻璃的小颗粒。在构建周期内，用于在其上进行构造的并重新定位的平台，每次降低一个单层厚度的高度。该过程重复，直到构建或模型完成。与 SLA 技术不同，这个技术过程不需要额外支撑，因为构造过程中未烧结的材料可以提供支撑。

Keys to Exercises of the Text

Ⅱ. 1. f 2. e 3. a 4. b 5. c 6. d 7. h 8. g

Ⅲ.

1. 现在的每个人都知道 3D 打印这个术语，但很多人谈论它时，实际上指的是 7 种增材制造过程之一。

2. 每个增材制造过程因使用的材料和机器技术而各不相同。

3. 您进行 3D 打印之前首先要准备一个 3D 数字模型。

4. "3D 打印"这个术语的最初含义是关于这样一个过程：用喷墨打印机头在粉末床上沉积一层一层的黏接材料。

Appendix C New Words List
（生词表）

a programmer's language   编程语言
aborigine   [ˌæbəˈridʒiniː]   n. 土著，土著居民
Aborigine   n. 澳大利亚土著居民
abound   [əˈbaʊnd]   vi. 富于；充满
AC（Alternate Current）   交流
academic   [ˌækəˈdemik]   adj. 学院的，理论的
accelerate   [əkˈseləreit]   vt. 使……加快；vi. 加速；促进；增加
access   [ˈækses]   vt. 使用；存取；接近   n. 进入；使用权；通路
accommodate   [əˈkɒmədeɪt]   vt. 供应；容纳；使适应；调解   vi. 适应；调解
account   [əˈkaʊnt]   n. 账目，理由；账户   vi. 解释
achievement   [əˈtʃiːvmənt]   n. 完成，达到；成就，成绩
acoustics   [əˈkuːstɪks]   n. 声学，音响效果，音质
acquisition   [ˌækwɪˈzɪʃ(ə)n]   n. 获得物，获得
activate   [ˈæktiveit]   vt. 刺激；使活动
actuator   [ˈæktjʊeɪtə]   n. 执行机构；激励者；促动器
adapt to   使适应于
adaptive modulation   自适应调制
Additive Manufacturing   增材制造
additive   [ˈædɪtɪv]   adj. 附加的；[数] 加法的   n. 添加剂，添加物
address   [əˈdres]   vt. 演说；写地址；向……致辞；提出；处理   n. 地址；演讲；致辞
advent   [ˈædvənt]   n. 出现，到来
advisor   [ədˈvaɪzə]   n. 顾问，指导教师，劝告者
advocate   [ˈædvəkeit]   vt. 提倡，主张，拥护   n. 提倡者；支持者；律师
afield   [əˈfiːld]   adv. 在战场上；去野外；在远处；远离
aggregate   [ˈægrɪɡət]   vt. 使聚集，使积聚；总计达   n. 合计   adj. 总机的
algorithm   [ˈælɡərɪð(ə)m]   n. 运算法则，演算法，计算程序
alter   [ˈɔːltə]   vt. 改变，更改
alternating voltage   交流电压
alternative   [ɔːlˈtɜːnətɪv]   adj. 选择性的；交替的   n. 二中择一；可供选择的办法、事物
Amateur and Citizens Radio Service   业余爱好者和市民无线电服务设备
Amazon.com.   亚马逊网站
ambitious   [æmˈbɪʃəs]   adj. 野心勃勃的；有雄心的；热望的；炫耀的
ampere   [ˈæmpɛə]   n. 安［培］
amplifier   [ˈæmplifaiə]   n. 放大器，扩大器；扩音器
amplitude   [ˈæmplitjuːd]   n. 振幅；广阔；丰富，充足幅度

analog	[ˈænəlɒg]	n. 类似物；模拟　adj. 有长短针的；模拟的

analog modulation　模拟调制

analogue	[ˈænəlɒg]	n. 相似物；adj. 模拟计算机的；类似的，相似物的
analyze	[ˈænəlaiz]	vt. 〈美〉分析；分解；解释；对……进行心理分析
and gate		和门；与门；与电路
animation	[ˌæniˈmeiʃn]	n. 动画，活跃；生气勃勃；兴奋
annotate	[ˈænəuteit]	v. 注释，评注
anode	[ˈænəud]	n. 阳极（电解），正极（原电池）
antenna	[ænˈtenə]	n. 天线；【动物学】触角，触须
anticipated	[ænˈtisəˌpeitid]	adj. 预期的，期望的；v. 预料，盼望
anticipation	[ænˌtisiˈpeiʃ(ə)n]	n. 希望；预感；先发制人；预支
antivirus	[ˈæntivairəs]	n. 抗病毒素；反病毒程序
applause	[əˈplɔːz]	n. 欢呼，喝彩；鼓掌欢迎
application	[ˌæpliˈkeiʃ(ə)n]	n. 应用；申请；应用程序；敷用

application program　应用程序

approach	[əˈprəutʃ]	n. 方法，途径；接近　vt. 接近；着手处理　vi. 靠近；走近
appropriate	[əˈprəupriət]	adj. 适当的；恰当的；合适的
architect	[ˈɑːkitekt]	n. 建筑师；缔造者
archive	[ˈɑːkaiv]	n. 档案馆；档案文件　vt. 把……存档
arduous	[ˈɑːdjuəs]	adj. 努力的；费力的；险峻的
arouse	[əˈrauz]	vt. 引起；唤醒；鼓励　vi. 激发；醒来
array	[əˈrei]	n. 数组，阵列；队列；排列；大批；衣服　vt. 排列，部署；打扮
ascertain (= find out) ascertain	[ˌæsəˈtein]	vt. 确定；查明；探知
assemble	[əˈsemb(ə)l]	vt. 集合，聚集；装配；收集　vi. 集合，聚集
assembly	[əˈsembli]	n. 集合，
assembly	[əˈsembli]	n. 装配；集会；集合；汇编

assembly language　汇编语言

assets	[ˈæsets]	n. 资产；有用的东西；有利条件；优点
associate	[əˈsəuʃieit]	v. 使联合，结交　adj. 副的
atlas	[ˈætləs]	n. 地图集
atomic	[əˈtɒmik]	adj. 原子的，原子能的；微粒子的
audacious	[ɔːˈdeiʃəs]	adj. 大胆的；鲁莽的；大胆创新的
automobile	[ˈɔːtəməbiːl]	n. 〈美〉汽车　vt. 驾驶汽车
automotive	[ˌɔːtəˈməutiv]	adj. 汽车的，机动车的
autonomous	[ɔːˈtɒnəməs]	adj. 独立的，自治的；自主的；自发的
available	[əˈveiləbl]	adj. 可用的；可获得的；有时间的
bandwidth	[ˈbændwidθ]	n. [通信] 频带宽度；[电子] [物] 带宽；
base	[beis]	n. 底部；基础；基极　adj. 低劣的　vt. 以……作基础

base unit　主机

base-emitter circuit　基-射回路

basis	[ˈbeisis]	n. 基础；（物体的）底部；基本原则
battery	[ˈbætəri]	n. 电池；一组，一套；炮台，炮位

battery-powered 电池驱动的
be divided into 被分成
be equivalent to 等于
be represented by 用……表示
be utilized for 用于
beam ['biːm] n. 梁；（光线的）束；电波 v. 播送
beep [biːp] vi. 嘟嘟响，n. 哔哔的声音，警笛声
bias ['baiəs] n. 偏见；斜纹；偏置 vt. 使存偏见 adj. 偏斜的
biased ['baiəst] adj. 有偏见的；结果偏倚的，有偏的
billing interface 计费接口
binary ['bainəri] adj. 二进制的；二元的；由两部分构成的 n. 二进制数
binary arithmetic 二进制算术
binder ['baində(r)] n. 黏结剂；包扎物，包扎工具；装订工
biology [bai'ɒlədʒi] n. 生物学；生物
biotechnology [ˌbaiə(ʊ)tek'nɒlədʒi] n. [生物] 生物技术；生物工艺学
bit [bit] n. 比特；一点；少量 [常与 a 连用，起副词作用]
blockade [blɔː'keid] n. 封锁；封锁部队；障碍物，阻碍物
blossom ['blɒsəm] vi. 发展；开花 n. （尤指果树的）花
breadboard ['bredbɔːd] n. 擀面板，案板，电路试验板
brightness ['braitnis] n. 亮度，明亮，光泽度，灯火通明，活泼，愉快
bring about 引起；使掉头
buddy ['bʌdi] n. 伙伴，好朋友 [美口] 密友 vi. 做好朋友，交朋友
bundled with 与……捆绑
burgeoning ['bɜːdʒəniŋ] adj. 迅速成长的；v. 迅速发展；发（芽）
buzzer ['bʌzə] n. 蜂鸣器；嗡嗡作声的东西；信号手
cable modem 光缆调制解调器
calculation [ˌkælkjuː'leiʃən] n. 计算；估计；计算的结果
calendar ['kælində] n. 日历
capability [ˌkeipə'biləti] n. 能力；容量；可能，可能性
capacitor [kə'pæsitə] n. 电容，电容器
capacity [kə'pæsiti] n. 能力；容量；资格，地位；生产力
capitalize ['kæpitəlaiz] vi. 利用，积累资本 vt. 使资本化；以大写字母写
capture ['kæptʃə] vt. 俘获，夺得 n. 战利品，俘虏
cast [kɑːst] vt. 浇铸；投，抛；计算；投射（光、影、视线等）
categorize ['kætəgəraiz] vt. 分类
cater ['keitə(r)] vt. 投合，迎合；满足需要；提供饮食及服务
cathode ['kæθəud] n. 阴极，负极
cathode-ray tube (CRT) 阴极射线显像管
cell phone (=cellular phone) 移动电话
cellular ['seljulə] adj. 细胞的；多孔的；由细胞组成的 n. 移动电话；单元
ceramic [si'ræmik] adj. 陶瓷的；陶器的；n. 陶瓷；陶瓷制品
chancellor ['tʃɑːnsələ(r)] n. 总理（德、奥等的）；（英）大臣；校长

charge　［tʃɑːdʒ］　*n.* 费用；负载；电荷　*vt.* 使充电　*vi.* 索价；充电
checkbook　［ˈtʃekbuk］　*n.* 支票簿；核算
chemical composition　化学成分，化学组成
chip　［tʃip］　*vt.* 削，凿；削成碎片　*vi.* 剥落；碎裂　*n.* 电子芯片；碎片
circuit　［ˈsəːkit］　*n.* 电路
cityscape　［ˈsitiskeip］　*n.* 都市风景
classify　［ˈklæsifaɪ］　*vt.* 分类，归类；分等；把……列为密件
Code Division Multiple Access（CDMA）　码分多址
coder　［ˈkəudə］　*n.* 编码器；编码员
coin　［kɔin］　*vt.* 铸造（货币）；杜撰，创造　*n.* 硬币，钱币
collaboration　［kəlæbəˈreɪʃn］　*n.* 合作；
collector　［kəˈlektə］　*n.* 收藏家；收税员；征收者；集电极
collector-emitter circuit　集－射回路
combination　［ˌkɑmbiˈneɪʃən］　*n.* 结合，联合，合并；联合体；密码组合
come about　发生；产生；改变方向
commonsense　［ˈkɑmənˈsɛns］　*adj.* 常识的，具有常识的
communication system　通信系统
comparable　［ˈkɑmpərəbl］　*adj.* 可比较的，比得上的
compartmentalization　［ˈkɔmpɑːtˌmentəlaiˈzeiʃən］　*n.* 区分，划分
compatible　［kəmˈpætɪb(ə)l］　*adj.* 兼容的
compensate　［ˈkɔmpenseit］　*v.* 偿还，补偿，付报酬
compete　［kəmˈpiːt］　*vi.* 比赛，竞争
competitive　［kəmˈpɛtitiv］　*adj.* 竞争的，比赛的；（价格等）有竞争力的
compile　［kəmˈpail］　*vt.* 编译；编制；编辑
complex　［ˈkɒmpleks］　*adj.* 复杂的；合成的　*n.* 复合体；情结；不正常的忧虑
complexity　［kəmˈpleksəti］　*n.* 复杂性，复杂的事物；复合物
complicated　［ˈkɔmplikeitid］　*adj.* 难懂的，复杂的
compound　［ˈkɔmpaund］　*vt.* 混合；合成　*n.* 化合物；混合物
comprehensive　［kɒmprɪˈhensɪv］　*adj.* 综合的；广泛的；有理解力的
computation　［ˌkɒmpjuˈteiʃən］　*n.* 计算，估计
conceive　［kənˈsiːv］　*vt.* 想到；想象；以为
conceptual　［kənˈseptjuəl］　*adj.* 概念上的
concrete　［ˈkɒŋkriːt］　*adj.* 混凝土的；实在的，具体的　*n.* 混凝土
conduct　［ˈkɒndʌkt］　*vi.* 带领；导电　*vt.* 管理　*n.* 行为；实施
conductivity　［kɒndʌkˈtɪvɪtɪ］　*n.* 导电性；传导性
configuration　［kənfigəˈreiʃn］　*n.* 布局，构造；配置；排列；结构；外形；组态
configure　［kənˈfigə］　*vt.* 配置，设置，按特定形式装配，安装
confine　［kənˈfain］　*vt.* 限制；禁闭　*n.* 界限，边界
connection　［kəˈnekʃn］　*n.* 连接；联系，关系；连接点
connectivity　［kɒnekˈtɪvɪtɪ］　*n.* 连接性；连通性可连接性
consistent　［kənˈsɪstənt］　*adj.* 一致的，一贯的，相容的，调和的
consolidate　［kənˈsɒlideɪt］　*vt.* 巩固，使固定；联合　*vi.* 巩固，加强

consortium　[kən'sɔːtɪəm]　n. 财团；联合；合伙
constant　['kɒnst(ə)nt]　adj. 不变的，恒定的，经常的　n. [数] 常数，恒量
constitute　['kɒnstɪtjuːt]　vt. 组成，构成；任命；建立
consultancy　[kən'sʌlt(ə)nsɪ]　n. 咨询公司；顾问工作
contain　[kən'teɪn]　vt. 包含，容纳；容忍；克制，遏制；牵制
contour　['kɒntʊə]　n. 轮廓；等高线；周线；概要　vt. 画轮廓；画等高线
Contour Crafting　轮廓工艺
conventional　[kən'venʃ(ə)n(ə)l]　adj. 传统的；常见的；惯例的
convergence　[kən'vɜːdʒəns]　n. 集中，收敛
convert　[kən'vɜːt]　vt. 使转变；使改变信仰；改建
convey　[kən'veɪ]　v. 传达，运输，转移；输送
cooperative　[kəʊ'ɒpərətɪv]　adj. 合作的；协助的；共同的
coordinate　[kəʊ'ɔːdɪnɪt]　v. 协调，综合　n. 同位格　adj. 同等的
coulomb　['kuːlɒm]　n. 库［仑］
counterfeit　['kaʊntəfɪt；-fiːt]　vt. 假装，伪装；n. 伪造品；adj. 伪造的
countersign　['kaʊntəsaɪn]　n. 口令；副署　vt. 副署；会签；确认
coverage　['kʌvərɪdʒ]　n. 覆盖，覆盖范围；新闻报道
Craig Mc Caw：董事长克莱格·麦科考
critical　['krɪtɪk(ə)l]　adj. 鉴定的；爱挑剔的；决定性的；评论的
cross-section　剖面
culmination　[kʌlmɪ'neɪʃ(ə)n]　n. 顶点；高潮
curly brace　花括号，大括号
current　['kʌrənt]　adj. 现在的；通用的；最近的　n.（水、气、电）流；涌流
customize　['kʌstəmaɪz]　vt. 定做，按客户具体要求制造
cyber　['saɪbə]　adj. 计算机（网络）的，信息技术的
cyber-physical system　信息物理系统
DC（Direct Current）　直流
de facto　[diː'fæktəʊ]　adj. 事实上（的），实际上（的）
debug　[diː'bʌg]　vt. 调试　n. [计] 调试工具
decibel　['desɪbel]　n. 分贝
decoder　[ˌdiː'kəʊdə]　n. 解码器，译码器；译码员
default　[dɪ'fɔːlt]　n. 默认（值），缺省（值），常用
defect　['diːfekt, dɪ'fekt]　n. 缺点，缺陷　vi. 叛变；变节
deficiency　[dɪ'fɪʃənsɪ]　n. 缺乏；不足的数额；缺陷，缺点
deflect　[dɪ'flekt]　vt. 使偏斜；使转向；使弯曲　vi. 偏斜；转向
delivery　[dɪ'lɪv(ə)rɪ]　n. 交付；分娩；递送
demodulation　[ˌdiːmɒdjʊ'leɪʃn]　n. 检波；反调制；解调制
dental　['dent(ə)l]　adj. 牙科的；牙齿的，牙的
deploy　[dɪ'plɔɪ]　vt. 配置；展开；使疏开　vi. 部署；展开　n. 部署
deposition　[ˌdepə'zɪʃ(ə)n, diː-]　n. 沉积物；矿床；革职
desktop　['desktɒp]　n. 桌面；台式机
desktop computer　台式计算机

detach [diˈtætʃ] vt. 分离；派遣；使超然
detect [dɪˈtekt] vt. 察觉；发现；探测
detection [diˈtekʃən] n. 察觉；侦查，探测；发觉，发现
device [diˈvais] n. 仪器；策略；商标图案
diagnostic [daɪəgˈnɒstɪk] adj. 诊断的；特征的 n. 诊断法；诊断结论
diagram [ˈdaɪəgræm] n. 图表，图解，示意图，线图；vt. 用图表示；图解
dial [ˈdaɪəl] n. 转盘；刻度盘；钟面 vi. 拨号
diamond [ˈdaɪəmənd] n. 钻石，金刚石；菱形；adj. 金刚钻的
diesel engine n. 柴油机
digital [ˈdidʒitəl] adj. 数字的；手指的 n. 数字；键
digital modulation 数字调制
diode [ˈdaɪəud] n. 二极管
Direct Digital Manufacturing 直接数字制造
disc [disk] n. 磁盘；唱片；圆盘
discrete [disˈkriːt] adj. 离散的 n. 分立元件；独立部件
display [dɪˈspleɪ] vt. 显示；表现；陈列 adj. 展览的 n. 显示（器）；炫耀
dissipation [ˌdisiˈpeiʃn] n. （物质、精力逐渐的）消散，分散
dissolve [diˈzɔlv] n. 渐渐消隐，溶化 vt. 叠化，叠化画面
distinctive [dɪˈstɪŋ(k)tɪv] adj. 有特色的，与众不同的
distinguish [disˈtɪŋgwɪʃ] vt. 区别，辨别
distinguishing [dɪˈstɪŋgwɪʃɪŋ] adj. 有区别的；v. 区别，表现突出
distribute [dɪˈstrɪbjuːt; ˈdɪstrɪbjuːt] vt. 分配，散布，分开，把…分类
diverse [daɪˈvɜːs; ˈdaɪvɜːs] adj. 多种多样的，不同的；变化多的
donor [ˈdəunə(r)] n. 捐赠者
doping [ˈdəupiŋ] n. （半导体）掺杂质，加添加剂；涂上航空涂料
downside [ˈdaunsaɪd] n. 下降趋势；负面，消极面 adj. 底侧的
downstream [ˌdaunˈstriːm] adv. 下游地；顺流而下 adj. 下游的
dramatically [drəˈmætɪkəlɪ] adv. 引人注目地，戏剧地，显著地，剧烈地
drawback [ˈdrɔːbæk] n. 缺点，不利条件；退税
drool [druːl] u. 流口水；说昏话
dual-trace oscilloscope 双踪示波器
DVB 数字视频广播
dynamic [daɪˈnæmɪk] adj. 动态的；动力学的；有活力的 n. 动态；动力
election [iˈlekʃən] n. （投票）选举，推举；当选；选择权
electromagnetic [ɪˌlektrə(ʊ)mægˈnetɪk] adj. 电磁的；电磁学的
electromotive force 电动势
electronic component 电子元件
electronic gadgets 数码产品
element [ˈelimənt] n. 元素，元件
emanate [ˈeməneɪt] vi. 发出，散发，发源；vt. 放射，发散
embed [imˈbed] vt. 栽种；使嵌入，使插入；使深留脑中；嵌入
embedded [ɪmˈbedɪd] v. 嵌入；adj. 嵌入式的，植入的，内含的

embedded module　嵌入模块
embrace　[ɪmˈbreɪs; em-]　v. 拥抱，包括
emitter　[iˈmitə]　n. 发射器，发射体；发射极
enable to　使能够
encode　[ɪnˈkəud]　n. 编码　vt. 把……编码
encompass　[ɪnˈkʌmpəs; en-]　vt. 包含；包围，环绕；完成
encrypt　[enˈkrɪpt]　v. 加密，将……译成密码
encyclopedia　[ɪnˌsaɪkləˈpiːdiə]　n. 百科全书
encyclopedias　[ˌensaikləuˈpiːdjəs]　adj. 如百科辞典的，百科全书式的
end-to-end　端对端；首尾相连
energy-efficient　能效高的；高能效的
enhance　[ɪnˈhɑːns]　vt. 提高；加强；增加；增强
entire　[ɪnˈtaɪə; en-]　adj. 全部的，整个的，全体的
envision　[ɪnˈvɪʒn]　vt. 想象；预想
equalizer　[ˈiːkwəlaizə]　n. 平衡装置；均衡器，平衡器；使相等的东西
equator　[ɪˈkweɪtə]　n. 赤道
equipment　[iˈkwipmənt]　n. 设备，装备；器材，配件
etched　[ˈetʃid]　adj. 被侵蚀的，风化的；v. 蚀刻（etch 的过去分词）
evolution　[ˌiːvəˈluːʃ(ə)n; ev-]　n. 演变；进化论；进展
evolve　[ɪˈvɒlv]　vt. 发展，进化；推断出　vi. 发展，进化；逐步形成
excess　[ikˈses, ˈekses]　a. 过量的；附加的　n. 超过，超越，过度，过量
execute　[ˈeksikjuːt]　vt. 实行；执行；处死
executive　[ɪgˈzekjutɪv]　adj. 行政的；经营的；执行的　n. 总经理；执行者
exploration　[ˌekspləˈreiʃ(ə)n]　n. 探测；探究；踏勘
extract　[ikˈstrækt, ˈekstrækt]　vt. 摘录；输出；取出　n. 摘录
fabricate　[ˈfæbrɪkeɪt]　vt. 制造；伪造；装配
facilitate　[fəsiliteit]　vt. 促进；帮助；使容易
feature　[ˈfiːtʃə]　n. 特点　vt. 以……为特色　vi. 起重要作用
femtocell　[ˈfemtəusel]　n. 家庭基站
fiber　[ˈfaibə]　n. 纤维；光纤（等于 fibre）
filament　[ˈfɪləm(ə)nt]　n. 灯丝，细线，单纤维；灯丝；细丝
filter　[ˈfɪltə]　n. 滤波器；过滤器；筛选　vt. 过滤，渗透　vi. 滤过，渗入
fingerprint identification　指纹识别
flash memory　闪存
flexible　[ˈfleksibl]　adj. 柔韧的，易曲的，灵活的
flicker　[ˈflikə]　vi. 闪烁；摇曳；颤动　vt. 使闪烁　n. 闪烁；闪光
flip-flop　[ˈflipflɒp]　n. 触发器；啪嗒啪嗒的响声　vt. 使翻转；使突然转变
floppy drive　软盘驱动器
FM broadcast receiver　调频广播接收器
focus　[ˈfəukəs]　n. 焦点；焦距；中心　v. （使）聚焦　n. 焦点焦距
folder　[ˈfəuldə]　n. 文件夹，折叠器，折叠机
formulation　[ˌfɔːmjuˈleiʃən]　n. 构想，规划；公式化

forward　　　['fɔːwəd]　　*adj.* 早的；向前的　*adv.* 正向；向前地　*n.* 前锋
foster　　　['fɒstə]　　*vt.* 怀抱，抱有（希望等），心怀；培养；养育
freeware　　　['friːweə(r)]　　*n.* 免费软件
frequency　　　['friːkw(ə)nsɪ]　　*n.* 频率
front-end amplifier　　前置放大器
function　　　['fʌŋ(k)ʃ(ə)n]　　*n.* 功能；函数；职责　*vi.* 运行；起作用
functional　　　['fʌŋkʃənl]　　*adj.* 功能的
functioning　　　['fʌŋkʃənɪŋ]　　*v.* 起作用（function 的现在分词）；正常工作
fundamental　　　[fʌndə'ment(ə)l]　　*adj.* 基本的，根本的　*n.* 基本原则
fundamentally　　　[fʌndə'mentəlɪ]　　*adv.* 根本地，从根本上；基础的
fuse　　　[fjuːz]　　*n.* 熔丝，熔线；*vt.* 使融合，使熔化；*vi.* 融合，熔化
Fused Deposition Modelling　　熔融沉积造模
gain　　　[geɪn]　　*n.* 收获；增益；利润　*vt.* 获得；增加　*vi.* 获利；增加
gateway　　　['geɪtweɪ]　　*n.* 门；网关；方法；通道；途径
generation　　　[dʒenə'reɪʃ(ə)n]　　*n.* 一代人；代（约 30 年），时代
geology　　　[dʒɪ'ɒlədʒɪ]　　*n.* 地质学；（某地区）地质情况
germanium　　　[dʒɜː'meɪnɪəm]　　*n.* 锗；锗元素（32 号元素，符号 Ge）
globe-spanning communications network　　全球通信网络
graphics　　　['græfiks]　　*n.* [测] 制图学；制图法；图表算法
hall　　　[hɔːl]　　*n.* 过道，走廊；食堂；学生宿舍　*n.* （英）霍尔（人名）
hammer out　　打造
handoff　　　['hændɔːf]　　*n.* 切换；传送；手递手传球（美国橄榄球）
handset　　　['hæn(d)set]　　*n.* 手机，电话听筒
hard drive　　硬驱，硬盘驱动器
hardware　　　['hɑːdweə]　　*n.* 【计算机】硬件；五金器具；军事装备
Harvard　　　['hɑːvəd]　　*n.* 哈佛大学；哈佛大学学生
Harvard University　　哈佛大学
have awareness of　　了解到，意识到
HDTV　　高清电视
headset　　　['hedset]　　*n.* 耳机
hence　　　[hens]　　*adv.* 因此；今后
hi-fi sound systems　　高保真立体声音响系统
high-end　　　['haɪend]　　*adj.* 高端的；高档的
hint at　　暗示；对别人暗示……
horizontally　　　[hɒrɪ'zɒntlɪ]　　*adv.* 水平地
hostile　　　['hɒstaɪl]　　*n.* 敌对；*adj.* 敌对的，敌方的，怀敌意的
identical　　　[aɪ'dentɪk(ə)l]　　*adj.* 同一的；完全相同的　*n.* 完全相同的事物
identification　　　[aɪˌdentɪfɪ'keɪʃ(ə)n]　　*n.* 鉴定，识别；认同；身份证明
illegal　　　[ɪ'liːgl]　　*adj.* 非法的；违法的；违反规则的　*n.* 非法移民；间谍
illustrate　　　['ɪləstreɪt]　　*vt.* 阐明，举例说明，图解
imaginary　　　[ɪ'mædʒɪn(ə)rɪ]　　*adj.* 虚构的，假想的；想象的；虚数的
i-Mode　　日本 1999 年推出的世界第一个智能电话

209

英文	音标	词义
impedance	[imˈpiːdəns]	n. 阻抗；全电阻；输入阻抗
implant	[ɪmˈplɑːnt]	vt. 种植；嵌入 vi. 被移植 n. [医] 植入物；植入管
implement	[ˈimplimənt]	vt. 实施，执行；实现，使生效 n. 工具；手段
import	[imˈpɔːt; ˈim-]	n. 进口；输入 vt. 输入，进口 vi. 输入，进口
impose	[imˈpəuz]	vi. 利用；欺骗；施加影响 vt. 强加；征税；以……欺骗
impurity	[imˈpjuərəti]	n. 杂质；不纯；不洁
in a nutshell		简而言之
in line with		符合；与……一致
in parallel with		与……并联
in series with		和……串联
in the order of		约为，大约
in the range of		在……范围内
incorporate	[inˈkɔːpəreit]	adj. 合并的，一体化的 vt. & vi. 包含；吸收；合并
increment	[ˈinkrimənt]	n. 增加，增量
indexing	[ˈindeksiŋ]	n. 指数化；[机械学] 分度，转位
individual	[indiˈvidjuəl]	adj. 个人的；独特的；个别的 n. 个人；个体
inductor	[inˈdʌktə]	n. 感应器，电感；授职者；感应体；扼流圈
Industrie	[ˈindəstri]	n. 工业（德语）
industry	[ˈindəstri]	n. 工业；产业（经济词汇）；工业界
infrared	[infrəˈred]	adj. 红外线的；n. 红外线
inherent	[inˈhiər(ə)nt, -ˈher(ə)nt]	adj. 固有的，内在的，与生俱来的
initiative	[iˈniʃətiv]	n. 主动权；首创精神的主动行动 adj. 主动的；自发的
injection	[inˈdʒekʃ(ə)n]	n. 注射；注射剂；充血；射入轨道
innate	[iˈneit; ˈineit]	adj. 先天的；固有的；与生俱来的
insert	[inˈsəːt, ˈinsəːt]	vt. 插入；嵌入 n. 插入物
instantaneous	[inst(ə)nˈteiniəs]	adj. 瞬间的，即刻的，猝发的
instruction	[inˈstrʌkʃ(ə)n]	n. 指示；命令；教导；用法说明 adj. 说明用法的
insulator	[ˈinsjuleitə]	n. 绝缘体；从事绝缘工作的工人
integrate	[ˈintigreit]	vt. 使……完整；vi. 成为一体 adj. 整合的；完全的 n. 集成体
integration	[intiˈgreiʃ(ə)n]	n. 结合，整合，一体化
intellectual	[intəˈlektʃuəl]	adj. 智力的；聪明的；理智的 n. 知识分子
intensity	[inˈtensiti]	n. 强烈，强度，强烈；[电子] 亮度；紧张
interact	[intərˈækt]	v. 交互
interaction	[intərˈækʃ(ə)n]	n. 互动
interactive	[intərˈæktiv]	adj. 交互式的；相互作用的
interconnect	[intəkəˈnektid]	vt. 使互相连接 vi. 互相联系
interdisciplinary	[intəˈdisiplin(ə)ri]	adj. 各学科间的
interface	[ˈintəfeis]	n. 接口；界面；接触面
interface	[ˈintəfeis]	v.（使通过界面或接口）接合，连接；[计算机] 使联系
Internet of Things		物联网
interoperability	[ˈintərɒpərəˈbiləti]	n. 互通性，互操作性

intricate	[ˈɪntrɪkət]	adj. 错综复杂的；难理解的；曲折；盘错
inversion	[ɪnˈvɜːʃən]	n. 倒置；倒转；反向反转
invest	[ɪnˈvest]	vt. 投资；覆盖；授予；包围 vi. 投资，入股；花钱买
investigation	[ɪnˌvestɪˈgeɪʃ(ə)n]	n. 调查，科学研究，学术研究调查
ionosphere	[aɪˈɒnəsfɪə]	n. 电离层
IR = infra-red	[ˌɪnfrəˈred]	adj. 红外线的
iris scanner	虹膜扫描仪	
ISP (Internet Service Provider)	网络服务提供者	
issuance	[ˈɪʃjuːəns]	n. 发布，发行
ITU	国际电信联盟	
jack	[dʒæk]	n. 插孔，插座，起重器 vt. （用起重器）抬起
jeweler	[ˈdʒuːələ]	n. 珠宝商；宝石匠
joystick	[ˈdʒɔɪstɪk]	n. 游戏操纵杆，控制杆
junction	[ˈdʒʌŋkʃən]	n. 连接，接合；交叉点；接合点
keep track of	跟踪，记录，保持联系	
kidney	[ˈkɪdni]	n. 肾，肾脏
kilovolt	[ˈkɪləʊvəʊlt]	n. 千伏 [特]
lag	[læg]	n. 落后；迟延；vt. 落后于 vi. 滞后；缓缓而行；蹒跚 adj. 最后的
lakeside	[ˈleɪksaɪd]	n. 湖边
landscape	[ˈlæn(d)skeɪp]	n. 风景，山水画，地形，美化
laptop	[ˈlæptɒp]	n. 膝上型轻便计算机，笔记本计算机
laptop computer	便携式计算机，手提计算机	
latency	[ˈleɪtənsi]	n. 潜伏时间；延迟时间
latitude	[ˈlætɪtjuːd]	n. 纬度
launch into	进入，投入	
layer	[ˈleɪə]	n. 层；【植物学】压条 vi. 分层堆积
light intensity	光强度	
light-emitting diode	发光二极管	
liver	[ˈlɪvə(r)]	n. 肝脏
logic gates	[计算机]	逻辑闸；逻辑门电路库；逻辑门
logically	[ˈlɒdʒɪkəli]	adv. 理论上，逻辑上
logistic	[ləˈdʒɪstɪk]	adj. 后勤学的；[数] 符号逻辑的
longitude	[ˈlɒndʒɪtjuːd; ˈlɒŋgɪ-]	n. 经度
lot size	批量	
LTE	长期演进技术	
lunar	[ˈluːnə]	adj. 月亮的，月球的；阴历的
machinist	[məˈʃiːnɪst]	n. 机械师
macroscopic	[ˌmækrə(ʊ)ˈskɒpɪk]	adj. 宏观的，肉眼可见的
magnitude	[ˈmægnɪtjuːd]	n. 数量，大小
mainframe	[ˈmeɪnfreɪm]	n. 大型主机
maintain	[meɪnˈteɪn]	vt. 维持，继续
manage	[ˈmænɪdʒ]	vi. 处理 vt. 管理，控制，操纵

manipulate [mə'nipjuleit] vt.（熟练地）操作，使用（机器等），巧妙地处理
manner ['mænə] n. 方式；习惯；种类；规矩；风俗
manufacture [mænju'fæktʃə] vt. 制造，生产；n. 制成品，产品，工业
marine [mə'riːn] adj. & n. 航海的，船舶的；舰队，海运业
mechanism ['mekənizəm] n. 机制；原理，程序，途径；进程；机械装置；技巧
megabytes [ˌmegə'baits] n. 兆字节，可简写为 MB
megohm ['megəum] n. 兆欧［姆］
memorandum [memə'rændəm] n. 备忘录；便笺
memory card 存储卡
meteorology ['miːtiə'rɔːlədʒi] n. 气象学
meter ['miːtə] n. 仪表；米 vt. 用仪表测量 vi. 用表计量
microampere ['maikrəu'æmpɛə] n. 微安
microcomputer ['maikrə(u)kɔmˌpjuːtə] n. 微计算机；［计］微型计算机
microprocessor [ˌmaikrəu'prəusesə] n. ［计］微处理器
microvolt ['maikrəuvəult] n. 微伏
millennium [mi'leniəm] n. 千年期，千禧年；一千年，千年纪念
milliampere [ˌmili'æmpɛə] n. 毫安
millivolt ['milivəult] n. 毫伏［特］
MIMO (Multi-input Multi-output) 多输入多输出技术
mind-numbing ['maindnʌmbiŋ] adj. 令人心烦意乱的；令人厌恶的；无法想象的
miniaturization [ˌminətʃərai'zeiʃn] n. 小型化，微型化
minority [mai'nɔːrəti] n. 少数民族；未成年 adj. 属于少数派的
MMS 多媒体短信服务
mobile communication network 移动通信网络
mobility [məu'biləti] n. 活动性，灵活性；迁移率，机动性
moderate ['mɔdəreit] adj. 适度的；中等的；温和的；有节制的；v. 节制；减轻
modify ['mɔdifai] vt. 更改，修改
modulate ['mɔdjuleit] v. 调整，转调
modulated carrier 调制载波，被调载波，已调载波
modulation [ˌmɔdjuˈleiʃən] n. ［电子］调制；调整
monitor ['mɔnitə] n. 监视器
monitoring ['mɔnitəriŋ] n. 监视，［自］监控，检验；v. 监视，监听，监督
monopoly [mə'nɔpəli] n. 垄断；专利权
morphology [mɔː'fɔlədʒi] n. ［生物］形态学、形态论
Multi-Jet Modelling 多重喷射造模
multimedia [mʌlti'miːdiə] n. 图文、视频和音频等的合成 adj. 多媒体的
multi-meter 万用电表，多量程仪表
multimeter [mʌl'timitə] n. 万用表，数字万用表，多用电表
multiple ['mʌltipl] adj. 多样的，多重的 n. 倍数，若干 v. 成倍增加
multiplicity [mʌlti'plisiti] n. 多样性
multiply ['mʌltiplai] v. 乘；（使）相乘 adv. 多样地 adj. 多层的
multisensory [ˌmʌlti'sensəri] adj. 多种感觉的；使用多种感觉器官的

nanotechnology [ˌnænə(ʊ)tek'nɒlədʒɪ] n. 纳米技术
narrow ['nærəʊ] adj. 狭窄的；勉强的；n. 海峡；隘路 vt. 使变狭窄 vi. 变窄
nature ['neɪtʃə] n. 自然；性质；种类；本性
navigate ['nævɪgeɪt] v. 航行，航海，航空，操纵，使通过
Navstar ['nævstɑː] n. 导航星（美国全球定位系统）
NAVSTAR 一种罗克韦尔公司制造的导航卫星
needle ['niːdl] n. 针；指针；针状物；刺激 vi. 缝纫；做针线
nest [nest] n. 巢，窝；安乐窝；温床 vt. 筑巢；嵌套 vi. 筑巢；找鸟巢
non-invasive 非插入式
nourish ['nʌrɪʃ] vt. 滋养，施肥于；抚养，教养；使健壮
novel material 新型材料
nozzle ['nɒz(ə)l] n. 喷嘴；管口；鼻
nuclear submarine n. 核潜艇
numerical [njuː'merɪkəl] adj. 数字的，用数表示的
obstruction [əb'strʌkʃn] n. 障碍物；阻碍物；阻碍；阻挠
OFDM 正交频分复用技术
ohm [əum] n. 欧[姆]
Ohm's Law 欧姆定律
Operating Systems：(OS) 操作系统
Operational Technology 经营技术；操作工艺
optical ['ɒptɪkəl] adj. 光学的；眼睛的，视觉的
optical detector 光检测器；光学探测器；光辐射探测器
optical fiber 光纤，光导纤维
optical signal 光信号
orbit ['ɔːbɪt] n. & v. 轨道，常轨，绕轨道而行
orbital ['ɔːbɪtəl] adj. 轨道的；眼窝的
organ ['ɔːgən] n. 机构；器官
organic [ɔː'gænɪk] adj. 有机的；(动、植物的) 器官的；有组织的
organize ['ɔːgənaɪz] vt. 组织；使有系统化
orient ['ɔːrɪənt; 'ɒr-] vt. 使适应；确定方向 n. 东方 adj. 东方的
oscilloscope [ɔ'sɪləskəup] n. 示波器
outer-ring 外环，外层，外圈，外包围圈
outwit [ˌaʊt'wɪt] vt. 瞒骗，以智取胜
Palm Pilot Palm 公司开发的一款掌上计算机
participant [pɑː'tɪsɪp(ə)nt] n. 参与者；关系者
patch [pætʃ] n. 片，补缀，碎片 vt. 修补；解决；掩饰
pattern ['pæt(ə)n] n. 模式；图案；样品
payroll ['peɪrəul] n. 工资单；发放的工资总额
peak-to-peak voltage 电压峰－峰值
perceptual [pə'septʃuəl] adj. 感知的；【心理学】知觉的
perform [pə'fɔːm] vt. 履行，执行 vi. 完成任务
performance [pə'fɔːm(ə)ns] n. 性能；绩效；表演；执行；表现

period ['pɪərɪəd] n. 周期，期间；课时 adj. 某一时代的
peripheral [pə'rɪfərəl] adj. 外围的；次要的；n. 外部设备，周边设备
permanent ['pɜːmənənt] adj. 永久的，永恒的 n. [口] 烫发
perpetrate ['pɜːpɪtreɪt] vt. 犯（罪）；做（恶）
personal computer systems 个人计算机系统
pertain [pə'teɪn, pɜː-] vi. 适合；属于；关于
pertain to 适合于
pertinent ['pɜːtɪnənt] adj. 相关的，相干的，中肯的，切题的
petaflop 千万亿次
phase shift 相位漂移/差别，移相
phone jack 听筒插口；听筒塞孔
photoconductor n. 光电导体；光电导元件；光敏电阻
photopolymer [ˌfəʊtəʊ'pɒlɪmə] n. 光聚合物，光敏聚合物；感光性树脂
phototransistor [ˌfəʊtəʊtræn'zɪstə] n. 光电晶体管
piconet 微微网
pilot ['paɪlət] adj. 试验性的；导向的；驾驶员的；辅助的
pioneer [ˌpaɪə'nɪə] n. 先驱者；[美国英语] 拓荒者 vt. 开拓
pitch [pɪtʃ] vt. 定调，定位于；掷 vi. 投掷；倾斜 n. 程度；音高；投掷
plaster ['plɑːstə(r)] n. 灰泥，涂墙泥；石膏；膏药
platform ['plætfɔːm] n. 平台；坛；讲台平台
polarity switch 极性开关
polymer ['pɒlɪmə] n. 聚合物
polymers ['pɒlɪməs] n. [高分子] 聚合物
popularity [ˌpɒpjʊ'lærətɪ] n. 普及，流行；名气；受大众欢迎
portable ['pɔːtəb(ə)l] adj. 手提的，便携式的；轻便的便携的 n. 便携式设备
positioning [pə'zɪʃnɪŋ] n. 定位；配置，布置 v. 定位；放置
potential difference 电位差
powder ['paʊdə] n. 粉；粉末 vt. 使成粉末；撒粉 vi. 搽粉；变成粉末
powerhouse ['paʊəhaʊs] n. 精力旺盛的人；发电所，动力室；强国
predictable [prɪ'dɪktəb(ə)l] adj. 可预言的
prediction [prɪ'dɪkʃən] n. 预言；预测；预报
prestore [priː'stɔː] v. 预存储
primarily ['praɪmərəlɪ] adv. 主要地，根本上；首先；最初的
prior to 在前，居先，先于
proactively [ˌprəʊ'æktɪvlɪ] adv. 主动地
procurement [prə'kjʊəmənt] n. 采购；获得，取得
product tagged 加标签的产品
productivity [ˌprɒdʌk'tɪvətɪ] n. 生产力；生产率；生产能力
proficient [prə'fɪʃnt] adj. 熟练的，精通的 n. 精通；专家，能手
programmable [ˌprəʊ'græməbl] adj. 可设计的，可编程的
projector [prə'dʒektə(r)] n. 电影放映机；幻灯机；投影
promising ['prɒmɪsɪŋ] adj. 有希望的，有前途的；v. 许诺，答应

propagation [ˌprɒpəˈgeɪʃən] n. 传播；繁殖；增殖
proportionate [prəˈpɔːʃənit] adj. 成比例的 vt. 使相称
prospect [ˈprɒspekt] n. 前景，期望，眺望处，景象；vt. 勘探，勘察
protocol [ˈprəutəkɒl] n. 草案，协议
prototype [ˈprəutətaɪp] n. 原型；标准，模范
provenance [ˈprɒvənəns] n. 出处，起源
pseudo [ˈsjuːdəu] adj. 假的，冒充的
publish [ˈpʌblɪʃ] vt. 出版；发表；公布 vi. 出版；发行；刊印
pulse [pʌls] n. 脉搏，脉冲 vt. 使跳动 vi. 跳动，脉跳
punch [pʌntʃ] v. 打孔 n. 冲压机；钻孔机 vt. 开洞；以拳重击 vi. 用拳猛击
QoS 服务质量技术
quadrillion [kwɔːˈdrɪljən] n. 千的五次方，百万的四次方；万亿
quiescent [kwaɪˈesənt] adj. 静止的；不活动的；沉寂的
radio frequency identification 射频识别
random [ˈrændəm] adj. 随机的；任意的
Rapid fabrication 快速制造
ratio [ˈreiʃiəu, -ʃəu] n. 比，比率
realm [relm] n. 领域，范围；王国
recipient [rɪˈsipiənt] n. 接受者；容器 adj. 接受的
reciprocal [rɪˈsiprəkəl] adj. 相互的；倒数的 n. 倒数
reconstitute [riːˈkɒnstɪtjuːt] vt. 再组成，再构成 vt. 重新设立
red-hot [ˈredˈhɒt] adj. 赤热的；激烈的，恼怒的；近期的，新的
redundancy [rɪˈdʌndənsi] n. 冗余；裁员
redundant [rɪˈdʌnd(ə)nt] adj. 多余的
referred to as 被称为
reflect [rɪˈflekt] vt. 反映，反射，表达，显示；vi. 映现，深思
refresh rate 刷新率
regulate [ˈregjuleit] vt. 调节，规定；有系统地管理；控制；校准
reliability [rɪˌlaɪəˈbɪləti] n. 信度；可靠性，可靠度
replicate [ˈreplɪkeɪt] vt. 复制，复写；[生] 复制
represent [reprɪˈzent] v. 表现，代表，体现，作为……的代表；描绘；回忆；再赠送
represent [reprɪˈzent] vt. 表现；代表；表示 vi. 提出异议；代表
research [rɪˈsɜːtʃ; ˈriːsɜːtʃ] n. 追究，探测，调查，探索；vi. 做研究
resistor [rɪˈzistə] n. 电阻器
resolution [ˌrezəˈluːʃən, -ˈljuː-] n. 分辨率；决议；解决
restriction [rɪˈstrɪkʃən] n. 限制，约束
resurrect [ˌrezəˈrekt] vt. 使复活；复兴；挖出 vi. 复活
reverse [rɪˈvɜːs] n. 背面；相反，反向 vt. 颠倒 adj. 反面的
revolution [ˌrevəˈluːʃn] n. 革命；彻底改变；旋转；运行，公转
rigid [ˈrɪdʒɪd] adj. 刚硬的，刚性的，严格的
rms voltage 电压有效值
robust [rə(ʊ)ˈbʌst] adj. 强健的，健康的，粗野的，粗鲁

rotation ［rəuˈteiʃən］ n. 【天文学】自转；旋转；轮流
router ［ˈruːtə（r）］ n. 路由器
routine ［ruːˈtiːn］ n. 常规，程序 adj. 日常的；例行的
satellite communication 卫星通信
scale ［skeil］ n. 刻度，衡量，数值范围 v. 依比例决定；攀登
SC-FDE 单载波频域均衡
sci-fi blockbusters 科幻大片（Sci-fi 是 Science-fiction 的缩写形式）
scramble ［ˈskræmbl］ vi. 攀登；仓促行动 vt. 攀登 n. 爬行，攀登
seamless ［ˈsiːmlis］ adj. 无缝的；不停顿的
Seattle ［siˈætl］ n. 西雅图（美国一港市）
security ［siˈkjuərəti］ n. 安全
segment ［ˈsegm（ə）nt］ n. 段；部分
selective ［sɪˈlektɪv］ adj. 精心选择的；选择的，不普遍的；淘汰的
Selective Laser Sintering 选择性激光烧结
self-powered 自供电的
semiconductor ［ˌsemikənˈdʌktə（r）］ n. ［电子］［物］半导体；半导体器件
sensor ［ˈsensə］ n. 传感器
sensor network 传感网
sequence ［ˈsiːkw（ə）ns］ n. 序列，继起的事，顺序
server 服务器
shareware ［ˈʃeəweə（r）］ n. 共享软件
short-cut ［ˈʃɔːtkʌt］ n. 捷径，近路 vi. 走捷径，抄近路
shuttle ［ˈʃʌtl］ n. 梭子；航天飞机 vt. & vi. 穿梭般来回移动
silicon ［ˈsilik（ə）n］ n. 【化学】硅；硅元素
simplify ［ˈsimplifai］ vt. 单一化，简单化
simultaneously ［ˌsimlˈteiniəsli］ adv. 同时地；
sintering ［ˈsɪntərɪŋ］ n. 烧结 v. 烧结；使熔结
sketch ［sketʃ］ n. 素描；略图；梗概
slice ［slais］ n. 薄片；菜刀 vt. 切下；将……切成薄片 vi. 切开
SNR（signal-to-noise ratio） 信噪比
software ［ˈsɔftweə］ n. 软件
solenoid ［ˈsəulənɒid］ n. 螺线管，螺线形电导管
sonic ［ˈsɔnik］ adj. 音速的；声音的；音波的
sophisticate ［səˈfɪstɪkeɪt］ n. 久经世故者，精通者 vt. 弄复杂 vi. 诡辩
sophisticated ［səˈfistikeitid］ adj. 富有经验的；精致的；复杂的；久经世故的
soundtrack ［ˈsaundtræk］ n. 声带；声迹；音轨；声道；电影配音
spam ［spæm］ n. 垃圾邮件
spark ［spɑːk］ n. 火花；闪光 vt. 发动；鼓舞 vi. 闪烁；发火花
speakerphone ［ˈspiːkəfəun］ n. 扬声电话
specific ［spiˈsifik］ adj. 特殊的，特定的；明确的；详细的
specification ［ˌspesifiˈkeiʃn］ n. 规格；说明书；详述
specify ［ˈspesifai］ vt. 指定；列举；详细说明；把……列入说明书

spectral efficiency　频谱效率
spectral　　['spektrəl]　　adj. ［光］光谱的
spectrum　　['spektrəm]　　n. 光谱；频谱；范围
spike　　[spaik]　　n. 长钉，尖峰信号，vt. 用尖物刺穿
sponsor　　['spɒnsə]　　n. 赞助者；主办者；保证人　vt. 赞助；发起
spontaneous　　[spɒn'teiniəs]　　adj. 自然的，不由自主的　adv. 自发地，天真地
spur　　[spɜː]　　n. 鼓舞，刺激　vi. 骑马疾驰；给予刺激　vt. 激励，鞭策
stack　　[stæk]　　vt. & vi. 堆成堆，垛；堆起来或覆盖住
standard　　['stændəd]　　n. 标准；度量衡标准　adj. 标准的
stands for　　代表；支持；象征；担任……的候选人
starch　　[stɑːtʃ]　　n. 淀粉
state-of-the-art　　最先进的；已经发展的；达到最高水准的
stationary　　['steiʃənəri]　　adj. 静止的，不动的；固定的；定居的
stem　　[stem]　　n. 干，茎；血统　vt. 阻止；vi. 阻止；起源于某事物；逆行
SteveBallmer　　史蒂夫·鲍尔默
storage　　['stɔːridʒ]　　n. 存储，仓库；贮藏所
strength　　[streŋθ]　　n. 力量；力气；兵力；长处
strip　　[strip]　　n. 带，条状；vt. 剥夺，剥去；
structure　　['strʌktʃə]　　n. 结构，构造
stuck　　[stʌk]　　v. 刺（stick 的过去式）adj. 动不了的；被卡住的
stunning　　['stʌnɪŋ]　　adj. 极好的；使人晕倒的；震耳欲聋的
subscriber　　[səb'skraɪbə(r)]　　n. 订户；用户；签署者；捐献者；捐助者
subsequent　　['sʌbsɪkw(ə)nt]　　adj. 后来的，随后的
subsequently　　['sʌbsikwəntli]　　adv. 随后，后来
subset　　['sʌbset]　　n. 子集
subsidiary　　[səb'sidiəri]　　adj. 辅助的，补充的；子公司
subtractive　　[səbːtræktɪv]　　adj. 减去的；负的；有负号的
successive　　[sək'sesɪv]　　adj. 连续的；继承的；依次的；接替的
superimpose　　['sjuːpərim'pəuz]　　vt. 重叠（安装，添加）
superimposed　　['sjuːpəim'pəuzd]　　adj. 叠加的；上叠的；重叠的；叠覆的，重叠的
amplification　　[ˌæmplɪfɪ'keɪʃnl]　　n. ［电子］放大（率）；扩大；详述
supplement　　['sʌplimənt]　　vt. 增补，补充
swap　　[swɒp]　　n. 交换；vt. 与……交换；以……作交换　vi. 交换；交易
switch on　　接通，开启
symbol　　['simbəl]　　n. 符号；象征
synchronize　　['siŋkrənaiz]　　vt. 使同步　vi. 同步；同时发生
Syncom satellite　　同步通信卫星
synthesize　　['sinθəsaiz]　　vt. 合成；综合　vi. 合成；综合
synthesizer　　['sinθisaizə]　　n. 音响合成器，综合器，合成器
system patch　　系统补丁
tag　　[tæg]　　n. 标签；vt. 尾随，紧随；连接；添饰　vi. 紧随
take advantage of　　利用

tap　　[tæp]　　vt. 轻敲；轻打；vi. 轻拍；n. 水龙头；轻打　vt. 采用；开发
tap out　　敲打出
technique　　[tek'ni:k]　　n. 技术，技巧，方法，表演法，手法
techniques　　[tek'ni:ks]　　n. 技巧，手法，技术，技能
telecommunications　　['telɪkəˌmju:nɪ'keɪʃənz]　　n. 无线电通信，电信；远程通信
templet　　['templɪt]　　n. 样板，模板（等于 template）
terminal　　['tɜ:mɪn(ə)l]　　n. 终端；末端；接线端；终点；终端机　adj. 末端的；终点的
terrestrial　　[tə'restrɪəl]　　adj. 地球的，地上的　n. 陆地生物；地球上的人
test leads　　表笔
texture　　['tekstʃə]　　n. 质地，肌理，结构，本质，实质
thebest-seller lists　　列入畅销书目录
the circuit element　　电路元件
the closed-switch state　　开关闭合状态
the electronic industry　　电子工业
the open-switch state　　开关断开状态
the Road Ahead　　未来之路；前方的道路
the video equipment　　视频设备
thermal　　['θɜ:m(ə)l]　　adj. 热的，保热的；温热的；n. 上升的热气流
thermoforming　　[θəmə'fɔ:mɪŋ]　　n. 热成型；热压成形
thermoplastic　　[θɜ:məʊ'plæstɪk]　　adj. 热塑性的　n.［塑料］热塑性塑料
Three-Dimensional Printing　　3D 打印
tissue　　['tɪʃu:; 'tɪsju:]　　n. 组织；纸巾
to account for　　解释，说明（原因等）；对……负责；占百分之
tossed around　　翻来覆去
track　　[træk]　　vt.（用望远镜、雷达等）跟踪　n. 足迹；轨道
traffic　　['træfɪk]　　n. 交通；通信量　vt. 用……作交换；在……通行　vi. 交易
transceiver　　[træn'si:və(r)]　　n. 无线电收发机，收发器
transfer　　[træns'fɜ:]　　n. 移动，传递，转移　vt. 转移，传递，转让
transistor　　[træn'zɪstə]　　n. 晶体管；晶体晶体管，半导体管
transmission　　[træns'mɪʃn]　　n. 变速器；传递；传送；播送
transmit　　[trænz'mɪt]　　vt. 传输；传播；发射；传达；vi. 传输；发射信号
transponder　　[træn'spɒndə]　　n. 应答器；转调器，变换器
triangle　　['traɪæŋg(ə)l]　　n. 三角形
trigger　　['trɪgə(r)]　　vt. 引发，引起；触发　vi. 松开扳柄　n. 扳机
trilateration　　[ˌtraɪlætə'reɪʃən]　　n.［测］三边测量（术）
trojan　　['trəʊdʒən]　　n. 特洛伊木马，木马程式
trouble-shooting　　n. 故障寻找
tuner　　['tju:nə]　　n.【无线电】（高频）调谐器，调谐设备；调音师
tweak　　[twi:k]　　vt. 稍稍调整（机器、系统等）
ultimate　　['ʌltɪmət]　　adj. & n. 最后的，最终的，根本的，最终
ultra large-scale integration（ULSI）　　超大规模集成电路
ultra　　['ʌltrə]　　adj. 极端的；过分的　n. 极端主义者；激进论者

ultrasound ['ʌltrəsaund] n. 超频率音响，超声波
UMB 超移动宽带
uneven [ʌn'iːvən] adj. 不均匀的；不平坦的；【数学】奇数的
UNIX n. UNIX 操作系统（一种多用户的计算机操作系统）
unparalleled [ʌn'pærəleld] adj. 无比的；无双的；空前未有的
unveil [ʌn'veil] vt. 使公之于众，揭开；揭 vi. 除去面纱；显露
update [ʌp'deit] vt. 更新；校正，修正 n. 现代化，更新
usability ['juːzəbləti] n. 合用，可用；可用性
valence ['veiləns] n. 价；原子价；化合价；效价
valve [vælv] n. 电子管，真空管，阀门；vt. 装阀门；以活门调节
variable ['vɛəriəbl] n. [数] 变数，变量 adj. 可变的，[数] 变量的
VCR control 录像机遥控器
version ['vɜːʃ(ə)n] n. 版本；译文；译本；倒转术
vertebral ['vɜːtibrəl] adj. 椎骨的；脊椎的
vertical ['vɜːtikəl] adj. 垂直的 n. 垂直线
Vessel ['vesəl] n. 脉管，血管
vibration [vai'breiʃ(ə)n] n. 振动，颤动，摇动，摆动，犹豫，心灵感应
video ['vidiəu] n. [电子] 视频；录像，录像机 adj. 视频的；录像的
videoconference ['vidiəu'kɒnfərəns] n. 视频会议
virtual ['vɜːtʃuəl] adj. 虚拟的；事实上的（但未在名义上或正式获得承认）
vision ['viʒ(ə)n] n. 视力；美景；眼力；想象力 vt. 想象；显现；梦见
visual ['viʒjuəl] adj. 看的，视觉的，形象的，视力的；栩栩如生的
visualization [ˌviʒjuəlai'zeiʃən] n. 形象化；清楚地呈现在心
voice traffic 未来语音流量；电话业务，语音信息量，语音通信量
VoIP (Voice over Internet Protocol) 互联网协议电话，IP电话
Web browsers 浏览器
WiMAX 全球微波互通存取
winding ['waindiŋ] n. 线圈，弯曲，缠绕物 adj. 弯曲的，蜿蜒的；卷绕的
worldwide ['wɜːl(d)waid] adj. 全世界的，世界范围的 adv. 在世界各地

219

Appendix D Widely Used Abbreviations for Technical Terms
专业术语常用缩略语

缩写	全称	释义
AA	Address Administration	地址管理
AC	Access Channel	接入通路/交流电
ACF	Authority Control Function	授权控制功能
A-D	Analog to Digital	模拟 – 数字转换
ADC	Analog to Digital Convertor	模拟/数字转换器
ADM	Adaptive Delta Modulation	自适应增量调制
ADPCM	Adaptive Differential Pulse Code Modulation	自适应差分脉冲编码调制
AI	Artificial Intelligence	人工智能
ALU	Arithmetic Logic Unit	算术逻辑单元
AM	Amplitude Modulation	调幅
APD	Avalanche Photodiode Detector	雪崩光电探测器
ASCII	American Standard Code for Information Interchange	美国信息交换标准码
AV	Audio Video	声视，视听
BC	Base-collector	集电结
BCD	Binary Coded Decimal	二进制编码的十进制数
BCR	Bi-directional Controlled Rectifier	双向晶闸管
BCR	Buffer Courtier Reset	缓冲计数器
BZ	Buzzer	蜂鸣器，蜂音器
C	Capacitance，Capacitor	电容量，电容器
CAD	Computer Aided Design	计算机辅助设计
CASE	Computer Aided Software Engineering	计算机辅助软件工程
CATV	Cable Television	电缆电视
CCD	Charge-Coupled Device	电荷耦合器件
CCTV	Closed-Circuit Television	闭路电视
CD	Compact Disc	光盘
CDMA	Code Division Multiple Access	码分多址
CDPD	Cellular Digital Packet Data	蜂窝数字分组数据
CPU	Central processing unit	中央处理器
CRTs	Cathode ray tubes	阴极射线管
D-AMPS	the Digital Advanced Mobile Phone System	数字移动电话系统
DAST	Direct Analog Store Technology	直接模拟存储技术
DIP	Dual In-line Package	双列直插封装
DP	Dial Pulse	拨号脉冲
DRAM	Dynamic Random Access Memory	动态随机存储器

DSP	Digital Signal Processor	数字信号处理器
DTL	Diode-Transistor Logic	数字电压表
DTV	Digital television	数字电视
DUT	Device Under Test	被测器件
DVM	Digital Voltmeter	数字仪表
ECG	Electro cardio graph	心电图
ECL	Emitter Coupled Logic	射极耦合逻辑
EDI	Electronic Data Interchange	电子数据交换
EIA	Electronic Industries Association	电子工业联合会
emf	Electromotive Force	电动势
EOC	End Of Conversion	转换结束
ESD	Electro-Static Discharge	场效应晶体管
EWB	Electronics Workbench	电子工作台
F/V	Frequency to Voltage Convertor	电压频率转换器
FET	Field-Effect Transistor	频率/电压转换
FFT	Fast Fourier Transform Algorithm	快速傅里叶变换算法
FTP	File Transfer Protocol	文件传输协议
FM	Frequency Modulation	调频
FSK	Frequency Shift Keying	频移键控
FSM	Field Strength Meter	场强计
FST	Fast Switching Transistor	快速晶闸管
FU	Fuse Unit	熔丝装置
FWD	Forward	正向的
GAL	Generic Array Logic	通用阵列逻辑
GND	Ground	接地,地线
GPRS	General Packet Radio Service	通用包无线电服务程序
GSM	Global System for Mobile Communication	全球数字移动电话系
GTO	Gate Turn Off Transistor	门极可关断晶体管
HCMOS	High Density COMS	高密度互补金属氧化物半导体(器件)
HDTV	High Definition Television	高清晰电视
HFHigh	Frequency	高频
HTLHigh	Threshold Logic	高阈值逻辑电路
HTML	Hypertext Markup Language	超文本链接标示语言
HTS	Heat Temperature Sensor	热温度传感器
ID	International Data	国际数据
IGBT	Insulated Gate Bipolar Transistor	绝缘栅双极型晶体管
IGFET	Insulated Gate Field Effect Transistor	绝缘栅场效应晶体管
IM	Instant Messaging	即时通信
I/O	Input/Output	输入/输出
IP	internet protocol	网际
IPM	Intelligent Power Module	智能功率模块
IPM	Incidental Phase Modulation	附带的相位调制

IR	Infrared Radiation/ Infra-red	红外辐射/红外线
IRQ	Interrupt Request	中断请求
ISDN	Integrated Services Digital Network	综合业务数字网
ISO	International Standards Organization	国际标准化组织
ISP	Internet Service Provider	网络服务提供者
I/V	Current to Voltage Convertor	电流/电压变换器
JFET	Junction Field Effect Transistor	结型场效应晶体管
LAN	Local area networks	局域网
LAS	Light Activated Switch	光敏开关
LASCS	Light Activated Silicon Controlled Switch	光控晶闸管开关
LCD	Liquid Crystal Display	液晶显示器
LDR	Light Dependent Resistor	光敏电阻
LED	Light-emitting diode	发光二极管
LRC	Longitudinal Redundancy Check	纵向冗余（码）校验
LSB	Least Significant Bit	最低有效位
LSI	Large Scale Integration	大规模集成电路
MCT	MOS Controlled Transistor	场控晶闸管
MCU	Micro Controller Unit	微控制器单元
MIC	Microphone	传声器
Min	Minute	分钟
MOS	Metal Oxide Semiconductor	金属氧化物半导体
MOSFET	Metal Oxide Semiconductor FET	金属氧化物半导体场效应晶体管
MPEG	Mention Pictures Experts Group	运动图像专家组
MSC	Mobile Switching Center	移动交换中心
MS-DOS	Microsoft Disc Operating System	微软磁盘操作系统
N	Negative	负
Ng	Not gate	非门
NTC	Negative Temperature Coefficient	负温度系数
OCB	Overload Circuit Breaker	过载断路器
OCS	Optical Communication System	光通信系统
OR	Type of Logic Circuit	或逻辑电路
OSI	Open Systems Interconnection	开放系统互联
OV	Over Voltage	过电压
PAM	Pulse Amplitude Modulation	脉冲幅度调制
PC	Pulse Code/ Personal Computer	脉冲码/个人计算机
PCB	Printed Circuit Board	印制电路板
PCM	Pulse Code Modulation	脉冲编码调制
PCS	Personal Communications Services	个人通信服务程序/业务
PDA	Personal Digital Assistant	个人数码助手，电子记事簿
PDCP	Personal Digital Cellular Packet	个人数字蜂窝分组
PFM	Pulse Frequency Modulation	脉冲频率调制
PG	Pulse Generator	脉冲发生器

PGM	Programmable	编程信号
PI	Proportional-Integral (controller)	比例积分（控制器）
PID	Proportional-Integral-Differential (controller)	比例积分微分（控制器）
PIN	Positive Intrinsic-Negative	光敏二极管
PIO	Parallel Input Output	并行输入输出
PLD	Phase-Locked Detector	同相检波
PLD	Phase-Locked Discriminator	锁相解调器
PLL	Phase—Locked Loop	锁相环路
PMOS	P-channel Metal Oxide Semiconductor FET	P 沟道
PPM	Pulse Phase Modulation	脉冲相位调制
PRD	Piezoelectric Radiation Detector	热电辐射探测器
PROM	Programmable Read Only Memory	可编程只读存储器
PRT	Pulse Recurrent Time	脉冲周期时间
PRT	Platinum Resistance Thermometer	铂电阻温度计
PUT	Programmable Unijunction Transistor	可编程单结晶体管
PWM	Pulse Width Modulation	脉冲宽度调制
R	Resistance, Resistor	电阻，电阻器
RCT	Reverse Conducting Thyristor	逆导晶闸管
REF	Reference	参考，基准
REV	Reverse	反转
R/F	Radio Frequency	射频
RGB	Red/Green/Blue	红绿蓝
RMS	Root-mean-square	均方根值
ROM	Read Only Memory	只读存储器
RP	Resistance Potentiometer	电位器
RST	Reset	复位信号
RT	Resistor with inherent variability depende	热敏电阻
RTD	Resistance Temperature Detector	电阻温度传感器
RTL	Resistor Transistor Logic	电阻晶体管逻辑（电路）
RV	Resistor with inherent variability dependent on the voltage	压敏电阻器
SBS	Silicon Bi-directional Switch	硅双向开关，双向硅开关
SCR	Silicon Controlled Rectifier	晶闸管整流器
SCS	Safety Control Switch	安全控制开关
SCS	Silicon Controlled Switch	晶闸管开关
SCS	Speed Control System	速度控制系统
SCS	Supply Control System	电源控制系统
SG	Spark Gap	放电器
SIT	Static Induction Transformer	静电感应变压器
SITH	Static Induction Thyristor	静电感应晶闸管
SP	Shift Pulse	移位脉冲
SPI	Serial Peripheral Interface	串行外围接口

SR	Silicon Rectifier	硅整流器
SR	Sample Relay	取样继电器
SR	Saturable Reactor	饱和电抗器
SRAM	Static Random Access Memory	静态随机存储器
SSB	Single-sideband / Single sideband	单边带发射机；单边带通信
SSR	Solid. State Relay	固体继电器
SSR	Switching Select Repeater	中断器开关选择器
SSS	Silicon Symmetrical Switch	硅对称开关，双向晶闸管
SSW	Synchr0—Switch	同步开关
ST	Start，Starter	启动，启动器
STB	Strobe	闸门，选通脉冲
TACH	Tachometer	转速计，转速表
TCP	Transfer Control Protocol	传输控制协议
TCP/IP	Transmission Control Protocol/Internet Protocol	协议
TFT	Thin Film Transistor	薄膜晶体管
TN	Twisted Nematics	扭曲向列
TP	Temperature Probe	温度传感器
TRIAC	Triodes AC Switch	晶体管交流开关
TTL	Transistor-Transistor Logic	晶体管，晶体管逻辑
UART	Universal Asynchronous Receiver Transmitter	通用异步收发器
UMTS	Universal Mobile Telecommunications System	通用移动电信系统
UNIVAC	Universal Automatic Computer	通用自动计算机
VCO	Voltage Controlled Oscillator	压控振荡器
VD	Video Decoders	视频译码器
VDR	Voltage Dependent Resistor	压敏电阻
V/F	Voltage—to-Frequency	电压/频率转换
V/I	Voltage to Current Convertor	电压/电流变换器
VLSIC	Very Large Scale Integrated Circuits	超大规模积体电路
VM	Voltmeter	电压表
VOIP	Voice Over Internet Phone	网络语音电话业务
VOIP	Voice Over Internet Protocol	互联网协议电话
VOM	Electronic Voltmeter	电子伏特表
VS	Vacuum Switch	电子开关
WAN	Wide Area Networks	广域网
WAP	Wireless Application Protocol	无线电应用协议信号

参 考 文 献

[1] 蒋忠仁. 计算机应用专业英语 [M]. 重庆：重庆大学出版社，2006.
[2] 周伯清. 电子信息专业英语 [M]. 北京：科学出版社，2009.
[3] 刘小芹，刘骋. 电子与通信技术专业英语 [M]. 4版. 北京：人民邮电出版社，2014.
[4] 庄朝蓉. 电子信息专业英语 [M]. 2版. 北京：北京邮电大学出版社，2013.
[5] 江华圣. 电工电子专业英语 [M]. 2版. 北京：人民邮电出版社，2010.
[6] 严稽文，胡伟成. 电信英语 [M]. 北京：人民邮电出版社，2006.
[7] 汤滟，丁宁. 电子信息类专业英语 [M]. 西安：西安电子科技大学出版社2009.
[8] 别传爽. 机电专业英语 [M]. 北京：北京理工大学出版社，2010.
[9] 朱一纶. 电子技术专业英语 [M]. 4版. 北京：电子工业出版社，2015.
[10] 司建国. 职业英语交际手册 [M]. 北京：外语教学与研究出版社，2009.
[11] 徐存善. 机电专业英语 [M]. 2版. 北京：机械工业出版社，2012.
[12] 潘超群. 电信专业常用英文缩略语词典 [M]. 北京：电子工业出版社，2006.
[13] 赵中颖. 职业英语——计算机类 [M]. 北京：机械工业出版社，2011.
[14] 马佐贤，戴金茂. 电子信息专业英语 [M]. 北京：化学工业出版社，2011.
[15] 呼枫. 机电专业英语 [M]. 北京：人民邮电出版社，2016.
[16] 徐存善. 电子与通信工程专业英语 [M]. 北京：机械工业出版社，2010.